PLANTS AND LITERATURE

CRITICAL PLANT STUDIES: PHILOSOPHY, LITERATURE, CULTURE

PLANTS AND LITERATURE
ESSAYS IN CRITICAL PLANT STUDIES

EDITED BY RANDY LAIST

Rodopi

Amsterdam - New York, NY 2013

Cover illustration by Jim Horwitz.

Cover design: Inge Baeten, Senti Media

The paper on which this book is printed meets the requirements of 'ISO 9706: 1994, Information and documentation - Paper for documents - Requirements for permanence'.

Le papier sur lequel le présent ouvrage est imprimé remplit les prescriptions de "ISO 9706:1994, Information et documentation - Papier pour documents - Prescriptions pour la permanence".

ISBN: 978-90-420-3748-9
E-BOOK ISBN: 978-94-012-0999-1

Printed in The Netherlands

Table of Contents

Acknowledgements

This book owes its existence to the vision, insight, and adventurousness of the leadership of the College English Association, who agreed to sponsor a panel in plant studies at the 2013 Modern Language Association conference. Kathrynn Seidler Engberg and her colleagues at the CEA were willing to take a chance on the possibility that plants have something to say that is worth listening to. I am grateful to the MLA for hosting what turned out to be a very animated panel discussion, and of course to the panelists, who provided such compelling approaches to the themes of critical plant studies.

This book is also deeply indebted to Michael Marder for his ground-breaking scholarship in the philosophy of vegetation and for his initiative in arranging with Rodopi to publish the series of books in critical plant studies of which *Plants and Literature* is the first volume. Dr. Marder's writing has challenged conventional attitudes about plants as well as about the very nature of consciousness. His influence on my own personal thinking about vegetation has been considerable, and the impact of his ideas can be discerned throughout all of the essays in this collection.

Of course, this book would not have been possible without the dedicated scholarship, penetrating intelligence, playful imagination, and rigorous professionalism of the contributing essayists. I thank you for the high quality of your contributions and I look forward to maintaining professional friendships that will endure now that this project is complete. I am also grateful to Christa Stevens and the other editors at Rodopi for all of their helpfulness.

I would also like to thank Goodwin College for supporting me as I have worked on compiling and editing this book, and I would like to express my eternal gratitude to my wife, Ann, and son, Tony, for both being perfect in every way.

Introduction

Randy Laist

Flower in the crannied wall,
I pluck you out of the crannies,
I hold you here, root and all, in my hand,
Little flower -but if I could understand
What you are, root and all, and all in all,
I should know what God and man is.
Alfred, Lord Tennyson

Plants play a vital role in the experience of being human. We eat them, we wear them, we inhabit structures built out of plant materials, and we use them for drugs, medicine, cosmetics, ornaments, and symbols. Twentieth-century archeologists organized human history around the inorganic objects that survived the centuries – the stone age, the iron age, the bronze age – but the more fundamental story of human progress is the development, diversification, and refinement of our ability to manipulate plants. Not only has agriculture always been the primary source of the bioenergy that has allowed human populations to balloon so prolifically, but the weaving of plants into baskets, the carving of trees into floating vessels, and, possibly, the use of plant-based psychotropic substances to provoke dream-visions have all played a crucial role in the emergence of modern globalized human beings. But plants also live among us as fellow species and, if the view of earth from outer space is any indication, as the dominant form of life on our planet. We are surrounded by them spatially, for the most part, in that land-plants and human beings tend to inhabit similar climates. Plants have played a pivotal role in the evolution of human anatomy. The shape of our hands and fingers are reverse-molds of millions of years of tree branches, even as the shape of our feet and pelvis has been shaped by millions of years of walking through savannah grasslands. Certain pre-modern Inuit are said to have lived in an exclusively seal-based culture and to have used no plants for food or any other purpose, but, of course, even this seemingly plant-less society would be entirely reliant on the phytoplankton populations at the base of the arctic food chain that make the existence of the seals possible. As the prophet Isaiah intuited before biophysical science confirmed it to be literally so, "All flesh is grass." Plants are the source of all life on earth, the principle life forms responsible for capturing light from the sun and storing energy in molecules of carbohydrates and other organic substances, the very stuff of which all life is composed. In Genesis, God creates plants on the third day, a full 48 hours before the creation of animals, in accordance with every contemporary third grader's understanding that plants were here long before animals. What the Bible does not mention is what every modern third grader also knows: that plants will still be here long after human beings are gone.

It is impossible to overstate the significance of plants to human life, and yet this simple fact is easily overlooked, taken for granted, or, perhaps, actively repressed in the semantic texture of urban, technological consciousness. Pre-modern societies celebrate their relationship with plants through cultural expressions such as harvest ceremonies, the use of plants for ritual purposes, and tree worship, and although many of these traditions have been translated into postmodern forms (Christmas, Halloween, Spring Break), the symbolic connection between the ritual and the plant tends to be drastically attenuated in such contemporary celebrations. Vegetation plays a crucial role in the formative myths of all cultures, from Yggdrassil, the World Tree of Norse lore, to Asvattha, the cosmic tree of the Upanishads, to the Tree of the Knowledge of Good and Evil and the Tree of Life in the Garden of Eden. When one scans contemporary culture for evidence of plant-based narratives, however, the most dramatic meta-phenomenon is the defoliation of the cultural imagination. This is a phenomenon that has arguably been underway since agriculture, by mastering a certain portion of the plant kingdom, allowed human beings to urbanize and, ironically, to stop having to think about plants all the time. Into the modern era, two hundred or even one hundred years ago, Wordsworth or Whitman could rely on sharing a botanical vocabulary with his readers – a familiarity with the appearance, properties, and geographical distribution of lilacs, yew trees, and hyacinths – that is not available to most modern readers. While it is the rare book or movie that does not include plants in some way, one has to dig pretty deeply into the B-movie bin to find narratives that are actually *about* plants, or in which plants play a central role. The situation is a little different with poetry, a genre which, in many traditions, has a deep symbolic interrelationship with flowering plants, and contemporary nature poets such as Gary Snyder and Mary Oliver fight a rearguard action against encroaching mental defoliation, but in poetry's most ubiquitous postmodern form, pop music, references to plants are largely limited to monocultures of roses, sugar, cocaine, and marijuana.

The scarcity of plant-life in the cultural canon of the contemporary West is particularly striking when contrasted against the ubiquity of stories that feature animals as characters, subjects, and symbols. Despite the fact that urbanization has taken human beings just as far away from animals as it has taken them from plants, the fewer animals there are in the wild, the more seem to crop up on television, in children's toys, in advertisements, and on YouTube. A sub-discipline of cultural studies, animal studies, has recently developed in an effort to better understand the complex ways in which our representations of animals influence our perception of what it means to be human. In 2009, the sub-discipline of animal studies became the focus of the Theories and Methodologies section of an issue of *PMLA*, an institutional affirmation of the validity of examining the representations of non-human agents in fiction and poetry. In her introductory guest column for that issue, Marianne DeKoven begins with a discussion of the puppy in *The Great Gatsby* that Tom buys for Myrtle, and her essay concludes that attentiveness to the subjectivity of animals "means that we humans can give up the burden of our solipsism and our reign over the planet and take our place among the animals in a posthuman conjuncture" (368). DeKoven's exclusive focus on the kingdom of animals, however, simply widens the solipsistic circle while maintaining an ecological tunnel-vision that disregards the plant species on which all animal life depends. Animal studies is essentially an extension of human studies; it is relatively easy to imagine the subjectivity of animals. Animals may be shaped differently than we or pursue a different mode of life, but the basic coordinates of human existence and animal existence are identical in many respects. Like other animals, we eat, drink, sleep, procreate, communicate, and ambulate. We are born and we die. As a result, DeKoven's various interpretations of the symbolic significance of Myrtle's puppy – it emblematizes Mytle's otherness as well as her subjective self, it is the baby that Tom and Myrtle will never have, its sickliness anticipates the death that will come to Myrtle herself, etc. – tend to echo the narrative dynamics established by the human characters. The puppy reflects in canine form what an attentive reader is already able to discern about the human characters.

When it comes to plants, however, we encounter a much more significant barrier to our imagination. Plants seem to inhabit a time-sense, a life cycle, a desire-structure, and a morphology that is so utterly alien that it is easy and even tempting to deny their status as animate organisms. Aristotle established the place of plants in Western philosophy in a famous passage from *On Animals* in which he describes plants as intermediaries in a continuum of sentience, with inanimate matter at one end and (what we still call, unjustifiably) "higher animals" at the other. "The whole genus of plants," Aristotle writes, "while it is devoid of life as compared to an animal, is endowed with life as compared with other corporeal entities. Indeed, as we just remarked, there is observed in plants a continuous scale of ascent toward the animal. So, in the sea, there are certain objects concerning which one would be at a loss to determine whether they be animal or vegetable" (6). According to the Aristotelean Chain of Being, plants inhabit a liminal zone between animate and inanimate entities. In Darwinian terms, of course, there is no basis for declaring that plants are less "advanced" than animals, but Aristotle's characterization of plants has been extremely influential in informing the manner in which plants have been represented in art and literature. On the one hand, Aristotle's relegation of plants to the bottommost rung of the ladder of life may be responsible for the rarity with which plants are treated as subjects or even as powerful objects in Western literature. At the same time, however, Aristotle's positioning of plants at the borderline between animate and inanimate registers charges the vegetable kingdom with an uncanny ontological potency.

Conveniently, an example of this effect can be located in *The Great Gatsby*. Daisy fancifully compares Nick to a rose, an attempt to suggest the same kind of easy parallelism between Nick and a plant that DeKoven finds between Myrtle and her puppy, but Nick instantly deflates the conceit, assuring her that he is "not even faintly like a rose" (14). The ultimate truth of Nick's denial of any easy parallelism between plant life and human life becomes devastatingly apparent to

Gatsby in the final moments of his own life. Whereas Gatsby had existed throughout the novel as an utter personification of humanist aspiration, in the wake of Daisy's failure to call him on the telephone, plants disclose a vision of a dehumanized planet that radically undermines Gatsby's self-perception: "He must have looked up at an unfamiliar sky through frightening leaves and shivered as he found what a grotesque thing a rose is and how raw the sunlight was upon the scarcely created grass. A new world, material without being real, whose poor ghosts, breathing dreams like air, drifted fortuitously about … like that ashen, fantastic figure gliding toward him through the amorphous trees" (161). After all the blue lawns, cut flowers, and machine-sliced lemons, the final significance of plants in *The Great Gatsby* is the manner in which they disclose a level of existence that is completely depersonalized. For Gatsby, the rose, the grass, and the trees belong to a world of blind biological processes unfolding without order or purpose or meaning. Rather than reflecting the human characters, as in DeKoven's reading of Myrtle's dog, this instance of plant imagery constitutes a nihilistic or posthuman inversion the very concept of humanity. Gatsby's rose is more than just a symbol; reckoning with its material nature reveals the bi-directional nature of the Aristotelian Chain of Being. At the same time that mankind can ascend along the slope, it can slide down again, through the zombie-like no-man's land represented by plants and into the deindividuated soup of inanimate existence. Gatsby's (and, reciprocally, Nick's) horror in the displacement of the rose from a poetic frame of reference and into a domain of biochemical, evolutionary, cellular, organic, and chthonian forces extends to Daisy and Myrtle, twin faces of the Fitzgeraldian feminine, who both happen to be named after flowering plants. Women, sex, procreation, and by extension, life itself, earth itself, nature and terrestrial existence *in toto* are all included in Gatsby's reverie of terminal nihilism.

The contrast between Daisy's rose, a trite and frivolous symbol, and Gatsby's rose, a vector of a subversive perception that undermines the entire symbolic landscape of human enterprise, reveals an important feature of the phenomenology of plants. The underlying witchery of plants is that they are more than merely symbols, although they do of course have a robust symbolic potency. Many botanically-minded literary critics have provided excellent surveys of the symbolic significance of various species of plants and flowers in Shakespeare and other canonical writers. In fact, flowers and fruits are semantic in their being – they have evolved with the explicit purpose of sending messages to mammals, birds, insects and other animals that pollinate them and disperse their seeds. For pre-agricultural foragers, the conspicuous features of plants would have been very meaningful signs, packed with significant information about the time of year, the geography of a region, sources of water, or local fauna. It is only a small step from this practical mode of signification to the spiritual and cultural kind of symbolism that flowers have come to embody. Indeed, flowers were probably among the first physical objects in the life-world of prehistoric human beings to take on exclusively symbolic properties. The Victorians took the semantic quality of flowers to an extreme degree with the precisely codified Language of Flowers, according to which various arrangements of flowers were capable of expressing nuanced emotional sentiments. The consideration of the symbolic profile of individual plants is certainly an important aspect of plant studies, but, to my mind, the most interesting thing about plants is ontological, rather than semantic. More interesting than the question of what plants mean is the question of what they *are*. This is particularly the case in the contemporary period, when the inability of the typical urbanite to discriminate among different kinds of flowers or trees tends to lump the entire plant kingdom under a single perceptual category: a category of things that are alive like we are, but alive in a way that is utterly different, closed off from our capacity for empathy, omnipresent but unknown, seductive but unresponsive. As Michael Marder observes, a plant "turns out to be not only a what but also a who," an uncanny crossbreed that transects ontological boundaries and, as Gatsby discovers, challenges our basic assumptions about what it means to be a living thing.

The writers of the essays collected in *Plants and Literature* each address this question of the deep and complex meanings that writers and readers have discovered in the verdant mystery of our photosynthetic fellow-travelers. Plants and human culture are so deeply interwoven that the discipline of plant studies is essentially coextensive with human studies. Any comprehensive approach to the subject would require the kind of total history of human culture that this book is obviously not intended to provide. To borrow a pair of vegetal terms from Deleuze and Guattari, this book is arborescent in that it is loosely organized around a chronological development of plant imagery as it has been employed in the literary culture of the past two hundred years, but it is, more fundamentally, rhizomatic, providing a kaleidoscopic array of perspectives and case studies. In the same way that a plant typically appears as a single shape that is made up of a diverse arrangement of leaves and branches whose relationship to one another shifts as the plant is perceived from different angles, the essays in *Plants and Literature* constitute a shifting tracery of images, interpretations, and concepts that reflect the diversity and variety of the manner in which literary texts have conceptualized our relationship to the green world.

The essays in this collection focus specifically on the representation of plants in the contemporary moment and in the Western world, Western modernity being that mode of social organization whose peculiar (dis)relation with the natural world has been chiefly responsible for the current state of ecological crisis, although we anticipate that future studies will apply the perspective of plant studies to non-Western and pre-modern texts. Specifically, our collection begins with the romantic era in Europe and America, a period which brought renewed cultural attention to the subject of nature and during which many of our conventional attitudes concerning the relationship between human beings and the non-human world became codified. In this section, Betsy Winakur Tontiplaphol examines the manner in which romantic attitudes shape the symbolic value of Jane Austen's depiction of dietary culture in Mansfield Park, Lynne Feeley looks at the letters of antebellum farmers in the US South to investigate how attitudes about the tractability or unruliness of plants affect agricultural practices and are related to more global questions of white men's dominance over the rest of the world, and Akemi Yoshida considers how Christina Rossetti and her contemporaries adapted the perennial theme of "the forbidden fruit" in a way that both extends and reinterprets this fundamental image of the symbolically resonant plant product.

The era of Modernism marks a distinct reimagination of the relationship between human beings and the natural world. Though a close reading of Frank Norris's *The Pit*, Graham Culbertson sheds light on the manner in which, although wheat appears to have been domesticated as a fungible commodity, it still exerts an uncanny will of its own that overwhelms the agency of the novel's protagonist. In her reading of Susan Glaspell's play *The Verge*, Stacey Artman demonstrates that Glaspell employs plant imagery as a symbolic correlative for a wider experimental redefinition of subjectivity, femininity, and meaning itself. Ria Banerjee pursues this symbolic association between plants and femininity in her analysis of the role played by greenery in two masterpieces of modernism, Virginia Woolf's *Jacob's Room* and Djuna Barnes' *Nightwood.*

In the postmodern era, a deeper appreciation for the complex ecologies that bind together both human and non-human populations is responsible for fostering a more explicit consideration of the significance of plants to human life. Ubaraj Katawal's reading of Leslie Marmon Silko's *Ceremony* looks specifically at the redemptive potency associated with plants in that novel, Rhona Trauvich's essay on *Oryx and Crake* examines how Margaret Atwood reinvents the biblical story of the Garden of Eden for a world that has been reshaped by genetic engineering, and Charlotte Pylyser investigates the manner in which Flemish graphic novelist Brecht Evens uses plant imagery to supplement his artistic vision in his work *The Making Of.* The representation of plants in popular culture provides a compelling window into the anxieties and possibilities that are associated with plants in the contemporary imagination. Stepahie Lim's essay provides a useful survey of "killer plant narratives" that have found their way to the silver screen and focuses specifically *The Little Shop of Horrors*, one of the most popular narrative franchises to exploit this trope. Hindi Krinski performs a close examination of one of the most popular plant-based superheroes, Swamp Thing, tracing Alan Moore's development of the character into a complex ecological icon. Finally, in a wide-ranging essay that looks at pesticide advertisements and agricultural textbooks, Aubrey Streit Krug contemplates the possibility of a "queer" perspective on plant bodies. The range of perspectives brought to bear on the subject of plant life by these various authors and critics comprise a novel vision of interspecies interdependence and stimulate a revitalized sensitivity to the cultural and ecological significance of plants.

Works Cited

Aristotle, "The History of Animals." *The Animals Reader*. Ed., Linda Kalof and Amy Fitzgerald. Oxford: Berg, 2007. Print.

DeKoven, Marianne. "Guest Column: Why Animals Now?" *PMLA* 124.2 (2009). 361-369. Print.

Fitzgerald, F. Scott. *The Great Gatsby*. NY: Scribner, 1925. Print.

Marder, Michael. "If Peas Can Talk, Should We Eat Them?" *New York Times*. April 23 2012. Web. May 29 2013.

The Progress of Vegetation: Subversion and Vegetarianism in *Mansfield Park*

Betsy Winakur Tontiplaphol

Abstract: Over the course of Austen's novel, Fanny Price, a woman simultaneously plant-like and gardener-esque, transforms Mansfield Park—the seat of a family destabilized by the violence of the slave trade and, eventually, the carnal appetites of its spoiled children—with what might best be described as the passive promotion of vegetarian values. When considered alongside such commonplace terms as "cornfield" or "hayfield," the name "Mansfield" evokes the dystopian image of a plot where men, the most valuable of beasts, are cultivated for consumption, but Fanny adheres to both Enlightenment and Romantic vegetarian principles, quietly asserting the superiority of the vegetal over the carnal and, concomitantly, the centrality of plants to civilized living.

Jane Austen was not a vegetarian. "My mother desires me to tell you that I am a very good housekeeper," she writes in 1798 to her sister Cassandra, "which I have no reluctance in doing, because I really think it my peculiar excellence for this reason – I always take care to provide such things as please my own appetite, which I consider the chief merit in housekeeping." The "pleas[ing]" "things" that Austen "provide[d]" for the family table included a variety of meats – "I have had some ragout veal, and I mean to have some haricot mutton tomorrow. We are to kill a pig soon" – and as the "good housekeep[ing]" letter continues, she reports that a "Leicestershire sheep, sold to the butcher last week, weighed 27 lb. and ¼ per quarter" (13). Her correspondence as a whole is punctuated by similarly matter-of-fact references to animals destined for human consumption. "The chicken," she writes in 1811, "are all alive and fit for the table, but we save them for something grand" (119). A few years later, she remarks with evident gratification on the arrival of "[a] Hare & 4 Rabbits from [Godmersham] yesterday": "[W]e are stocked for nearly a week"

(183). In other descriptions of her "own appetite," Austen emphasizes the past over the future – that is, the veal she "[has] had" over the mutton she "means[s] to have." "[E]ating Turkies" is a "Christmas Duty" (125), she asserts, and in addition to surveying the contents of her homes' larders, her letters recount a number of meat-heavy meals, including "a good dinner, to which we sat down about lobster" (24), a supper of "very nice roast fowl" (153), and a "magnificent" repast at which "[w]e could not with the utmost exertion consume above the twentieth part of the beef" (48). If that final comment regarding the overabundance of beef encountered en route to Bath – "[b]etween Luggershall and Everley we made our grand meal" (48) – implies criticism, it would seem to be the criticism of a scrupulous economist, not an abstemious diner. Austen's correspondence also testifies to her general awareness of food prices, and as the beef-feast letter continues, she lists the costs of provisions in Bath: "I am not without hopes of tempting Mrs. Lloyd to settle in Bath; meat is only 8*d.* per pound, butter 12*d.*, and cheese 9½*d.* You must carefully conceal from her, however, the exorbitant price of fish: a salmon has been sold at 2*s.* 9*d.* per pound the whole fish" (49).

When Emma Woodhouse points out that she and Mr. Knightley are "not really so much brother and sister as to make it at all improper" for them to dance together, Mr. Knightley responds emphatically – "Brother and sister! no, indeed" (871) – and given the epistolary data, it seems reasonable to appropriate Mr. Knightley's unequivocal characterization of his relationship with Emma to describe Austen's own relationship to meat. Jane Austen a vegetarian? No, indeed. In truth, however, Emma's spontaneous observation speaks to her (and, arguably, Austen's) uneasiness regarding her sibling-esque bond with Mr. Knightley, whose brother has long been married to Emma's sister and who, upon his own marriage to Emma herself, agrees to reside at his bride's childhood home, where he has been a frequent visitor for Emma's entire life. I would suggest that Austen's remarks about meat hint at similarly sublimated anxieties. In the late eighteenth and early nineteenth centuries, vegetarianism was an increasingly visible dietetic-cum-ethical movement, and *Mansfield Park*, first published in 1814, debuted at roughly the same time as a variety of popular and influential vegetarian tracts, including Dr. William Lambe's *The Effects of a Peculiar Regimen* (1809), John Frank Newton's *The Return to Nature* (1811), and Percy Bysshe Shelley's *A Vindication of Natu-*

ral Diet (1813), which the poet appended to "his famous radical poem" *Queen Mab* (Stuart 372-375). At the age of ten, Fanny Price, the heroine of *Mansfield Park*, moves from her parents' modest home in Portsmouth to the estate named in the novel's title, and I aim to demonstrate that Austen's flora-loving protagonist functions, over the course of her story, to transform Mansfield – the seat of a family destabilized by the violence of the slave trade and, later, the carnal appetites of its spoiled children – with what might best be described as the passive promotion of vegetarian values. The Bertram family's disturbing reliance on animal foods and pleasures is called into question and, to some degree, reversed by the novel's conclusion; Fanny, who grows at and *into* Mansfield with the steadiness of transplanted ivy, quietly but effectively asserts the superiority of the vegetal over the carnal and, concomitantly, the centrality of plants to civilized living.

Although every item on the Mrs. Lloyd-inspired grocery inventory derives from animals, Austen's cognizance of *value* – meat, butter, and cheese are good values in Bath, while fish is not – signifies, I think, her attentiveness to contemporary food debates, in which English advocates of vegetarianism often cited the relative affordability of vegetable products in order to condemn the consumption of meat as unpardonable waste. In *The Bloodless Revolution: A Cultural History of Vegetarianism from 1600 to Modern Times*, Tristram Stuart summarizes the perspective of Dr. George Cheyne, "the most influential vegetarian in eighteenth-century Britain" and an important forerunner to the vegetarians publishing in the first decades of nineteenth century (163):

> Meat, considered by most of his contemporaries to be the most desirable and nourishing of all foods, Cheyne perceived as harbouring a richness verging on excess. The poor saved meager wages to buy the occasional joint of flesh; the rich consumed it in staggering quantities. In Britain especially, beef-eating was becoming synonymous with national pride and the annual consumption of 208 pounds of meat per head consumed by the British navy was seen as crucial to the nation's military prowess. Cheyne's admonition that people should avoid meat [...] flew in the face of common opinion. (173)

Despite her acknowledged taste for meat, then, Austen's comment about a spread of beef almost twenty times larger than necessary suggests a Cheynian discomfort with the "staggering quantities" of flesh typically consumed by Britain's wealthiest eaters. She never refers explicitly to Cheyne and his disciples, but it seems most likely that Austen knew their work, which addressed not only the economic ramifications of a carnivorous diet but also its medical consequences. Having established "a successful practice in London," Cheyne "set up shop in […] Bath," and although he had long since expired by the time Austen moved to Bath in 1801, his influence persisted in that health-conscious town, "where crowds flocked every year to drink mineral water and bathe in the therapeutic springs" (Stuart 167)[1]. Even more significant is the fact that Cheyne's most famous patient was Samuel Richardson, Austen's favorite novelist. "Over the years," Stuart asserts, "Cheyne taught Richardson about the link between diet and nervous sensibility; his ideas made their way into Richardson's novels and thus trickled into the mainstream of eighteenth-century emotional, literary and domestic culture" (181). Although Stuart acknowledges that Emma's father, the gruel-loving Mr. Woodhouse, embodies a biting parody of fastidious eating – "In accordance with Cheynian dietary principles, Woodhouse's 'conviction of suppers being very unwholesome' forces him to dissuade his guests from eating the food laid before them" (Stuart 193) – he also identifies in Austen's writing a quiet sympathy with other aspects of Cheyne's vegetarianism, rooted as it was in the belief that the consumption of meat generated "hydraulic blockages in the blood and nerves" (Stuart 166). "[F]or all [her] mockery," Stuart writes, "Austen's personal letters reveal that she was not immune to dietary doctrines herself when, for example, she betrays her assumption that full meals clog the brain: 'Composition seems to me Impossible, with a head full of Joints of Mutton'" (193). Meat-eating, in other words, can be costly in more ways than one; to value mental clarity as much as money is, Austen suggests, to eschew the "full"-ness that accompanies meaty meals.

In addition to addressing the nation's fiscal health and its citizens' physico-emotional wellness, vegetarians expressed concerns about England's morality. For Cheyne, "[m]eat built a wall – hardened by salts and oil – between man and his natural repulsion to flesh-eating";

---

[1] The letter which refers to the "grand meal" of beef was presumably begun in the morning of May 5, 1801. By that Tuesday night, Austen had walked with her uncle when he "went to take his second glass of water" (50). Austen's "own parents," Stuart writes, "had been subjected [to Cheynian dietary excesses] during their residence in Bath" (192).

Cheyne aimed "[to reconnect] man with his innate sympathy," which was, in his conception, "the body's inbuilt mechanism for spiritual salvation" (Stuart 178). The good doctor bullied Richardson in his letters, "warn[ing]" him

> that friends, family, and other doctors would try to get him to eat meat. [Cheyne] contemptuously lumped all those tempters together as "free-livers," "Beef-eaters," "cannibals," "Voluptuous and Flesh-eaters," likening them to carnivorous beasts eating "Dog's meat," and outrageously suggesting that they were in cahoots with the devil himself. Resisting them would be like re-enacting Christ's passion: Richardson was to respond with the prayer, "Father forgive them for they know not what they say or do." [...] The teetotal vegetable regimen was not just a cure, it was an identity, a lifestyle he called "living strictly aqueously and vegetably." (Stuart 185)

Richardson expressed doubts, but Cheyne's invectives about England's "carnivorous beasts" were, on the whole, effective. Richardson's virtuous heroines, Stuart points out, distinguish themselves dietetically from the "devil[s]" who persecute them: "[Pamela's and Clarissa's] foods of choice are bread, butter, water, tea, milk, toast, and chocolate. More often than not they refuse to eat anything at all, and signify their dissent from the tyrannical patriarchy that oppresses them by refusing to share their predominantly carnivorous meals" (186). Stuart continues, noting that

> Clarissa affectionately cares for her poultry and Pamela exhibits her sympathy for animals by releasing a carp and sowing the horse beans she has been using as fish bait in a flowerbed: "I will plant life," she tells her indignant carp-killing guardian, "while you are destroying it." [Clarissa's persecutor and, ultimately, rapist] Lovelace, by contrast, proudly acknowledges the continuity between his abuse of animals and women: "We begin with birds as boys, and as men go on to ladies; and both perhaps, in turns, experience our sportive cruelty." (189)

Mansfield Park's Fanny Price resembles Richardson's Clarissa in a variety of ways. A more demonstrably devout Christian than any other Austenian heroine, she prizes a cross pendant given to her by her brother William, and when she returns to her parents' slovenly home for a punishing visit, she pointedly abstains from meat. "I could not tell whether you would be for some meat, or only a dish of tea after your journey," Mrs. Price frets upon her arrival; William and Fanny "both [declare] they should prefer [tea] to anything" (633). Mrs. Price laments that "[i]t is very inconvenient to have no butcher in the street" (633), but during her exile in Portsmouth, Fanny survives on "biscuits and buns," unable to stomach the coarse food served "with [...] half-cleaned plates, and not half-cleaned knives and forks" at the Prices' table (653).

At the same time, however, Fanny's morality (and, along with it, Austen's vegetarian inspiration) is, at its core, at least as Romantic as it is eighteenth-century, as Shelleyan as it is Richardsonian or Cheynian. In his vital study of Austen and the British Romantic poets, William Deresiewicz denies Austen any affiliation with Shelley (2), the nineteenth century's best known literary vegetarian, but his own analysis of *Mansfield Park*'s ethos of "substitution" – "a set of psychic processes whereby individuals adjust to deprivation or loss by accepting alternative objects of desire" (57) – resonates profoundly with Shelley's Romantic vegetarianism. Shelley's principles derived from Cheyne's and those of his numerous inheritors, and although that circumstance alone offers, to my mind, enough reason to speculate about Austen's awareness of Shelley's writing, Deresiewicz's own reliance on eating metaphors to describe the "scarcity economics" that necessitate *Mansfield Park*'s numerous substitutions – "[W]hat one eats, the other cannot eat... This is life in the jungle [...] [a]nd the beasts must feed" (66) – hints at a link between Austen's project, as Deresiewicz intelligently theorizes it, and Shelley's radical verse. After all, at the most fundamental level, vegetarianism *is* substitution; proponents of a "vegetable regimen" advocate eating non-animal foods instead of meat, not abstaining from eating altogether. To eat "vegetably" is not to fast, and as Stuart notes, Shelley was unique in his contention that vegetarian living *corrected* for a long-standing but nonetheless spiritually damaging substitution: "[Shelley] proposed that 'the mythology of nearly all religions' testified that humans were originally vegetarian; that their consumption of meat precipitated global disjunction; and that one day they would re-establish harmony by returning to the natural diet. In the meantime, Shelley suggested, individuals could achieve this in their own domestic sphere" (377). To choose lettuce over lamb, then, is to *re*adjust psychically to humans' *original* "objects of desire," which, in more abstract terms, included peaceful – as opposed to violent – living. "Shelley merely pointed out that meat-eating was the most obvious source of cruelty in society," Stuart observes, and "[i]n a string of early poems" – including *Queen Mab* and *The Revolt of Islam* – "Shelley developed his fantasy that converting to vegetarianism would reverse malignancy throughout the world" (382, 383). Put more simply, we are what we eat, and Shelley's belief that consuming meat "'animalised' humans" (Stuart 397) accompanied a corresponding veneration of plants, the consumption of which, in addition to saving the lives of sentient beasts, could theoretically imbue vegetarians with a superhuman capacity for feeling. Although they don't experience pain, plants respond with what seems, at times, like magical sensitivity to the environments in which they grow, and Shelley explores that phenomenon in his strange neo-Platonic meditation "The Sensitive Plant" (1820), a poem that post-dates *Mansfield Park* but nonetheless offers some insight into the culture of Romantic vegetarianism that developed in the first decades of the nineteenth century. Cheyne associated vegetarianism with Christianity – through "a bizarre reinterpretation of the Bible," Cheyne disputed, in Christian terms, "the scriptural authority for eating meat" (Stuart 177-178) – but Shelley, religiously radical and quintessentially Romantic, maintained that the divine abides in nature, particularly its vegetal elements. Shelley's Sensitive Plant is a "companionless" organism in an "Eden"-like garden "[t]ended… from morn to even" by a goddess-esque "Spirit": "There was a Power in this sweet place/An Eve in this Eden; a ruling Grace/Which to the flowers, did they waken or dream,/Was as God is to the starry scheme" (I.12; II.1-4, 9, 17). The Sensitive Plant, however, possesses a "Spirit" of its own, and when the garden's protectress dies, it mourns – "The Sensitive Plant, like one forbid,/Wept, and the tears within each lid/Of its folded leave, which together grew,/Were changed to a blight of frozen glue" – and ultimately decays (III.78-81). Nevertheless, Shelley asserts, "[t]hat garden sweet, that lady fair,/And all sweet shapes and odours there,/In truth have never passed away" (III.130-132). In other words, the garden's holy ideals – "love, and beauty, and delight" (III.134) – persist in the face of organic "death [or] change" (III.135), and Fanny, despite her devotion to traditional Christianity, consistently expresses a Shelleyan commitment to the enduring and transporting "sublimity of Nature." Peering out a window at "the brilliancy of an unclouded night, and the contrast of the deep shade of the woods," Fanny "[speaks] her feelings" thus:

> Here's what may leave all Painting, and all music behind, and what Poetry only can attempt to describe. Here's what may tranquillize every care, and lift the heart to rapture! When I look out on such a night as this, I feel as if there could be neither wickedness nor sorrow in the world; and there certainly would be less of both if the sublimity of Nature were more attended to, and people were carried more out of themselves by contemplating such a scene. (487)

Deresiewicz's Romantic touchstone is Wordsworth, and I don't deny the Wordsworthian aspects of Fanny's psyche; Fanny's sitting room is adorned with little more than an image of Tintern Abbey, and there's much of Wordsworth in the foregoing celebration of nature's pleasures and powers. However, when Edmund follows Fanny's paean to nocturnal beauty with a remark about his cousin's exquisite "*taste* for nature" (487, my emphasis), his diction brings a more Shelleyan interest in diet to bear on Fanny's Romantic notions. Indeed, Fanny's assertion that there "would be less" suffering if "Nature were more attended to" echoes Stuart's account of Shelley's confidence in vegetarianism's capacity to "reverse malignancy throughout the world."

Fanny herself begins as a Sensitive Plant akin to that described in Shelley's poem: "companionless" and weepy but also, spiritually speaking, unexpectedly durable. The first chapter of *Mansfield Park* closes with Mrs. Price's "sanguine [...] hope" that her daughter, "somewhat delicate and puny," will soon grow "materially better for change of air" (429), and in subsequent scenes, Austen documents the fraught relocation – much later, Sir Thomas calls it a "transplantation" (577) – of her heroine, whom she consistently describes as more floral than faunal. Upon first encountering the Bertram family, Fanny seems the proverbial violet, "sweet" and "pretty" but "exceedingly timid and shy, and *shrinking* from notice" (430, emphasis added). The Miss Bertrams, "well-grown and forward for their age" (430), regard her as only minimally sentient, since in addition to seeming "little struck with the duet they were so good as to play," she "had never learnt French" (431). When granted a school holiday in order to become better acquainted with their bashful cousin, Maria and Julia soon abandon their half-hearted attempts to cultivate a relationship with Mansfield's shrinking violet, choosing instead to "[make] artificial flowers" (431). Cultivation falls to Edmund Bertram, who, following a chance encounter with a tearful Fanny, consciously exchanges benign neglect for something akin to nurturing: "He had never knowingly given her pain, but he now felt that she required more positive kindness" (432). Unlike his sisters, who seem put off by Fanny's plantishness, Edmund finds his cousin "an interesting object" (432), and his consideration facilitates her ultimately successful resettlement. Having been uprooted from familiar soil, Fanny slowly "learn[s] to transfer" – a term frequently associated, in the horticultural world, with repotting or otherwise relocating plants – "in [Mansfield's] favour much of her attachment to her former home" (434). Fanny's essential nature remains unchanged; "[k]ept back as she was by every body else, [Edmund's] single support could not bring her forward," could not *reconstitute* Fanny's essentially vegetal character despite her mother's most "sanguine" – here, I emphasize that word's etymologic ties to *blood* – wishes. As Mrs. Price had hoped, however, the "change of air" does engender growth, if not transmutation: "[Edmund's] attentions were [...] of the highest importance in assisting the improvement of her mind, and extending its pleasures… [H]e recommended the books which charmed her leisure hours, he encouraged her taste, and corrected her judgment" (435). Extending, encouraging, correcting: Edmund trains Fanny like a patient gardener trains a trellis vine, and if, over the course of the novel, Fanny becomes less reminiscent of the always low-lying violet, she remains a wallflower, an ivy-esque entity that creeps up and into the spaces, physical and social, that define her environment. Initially affiliated with discomfort – "[S]he crept about in constant terror of something or other," Austen writes of the earliest days (430) – Fanny's creeping finally secures her place at Mansfield. Consider, for example, her inch-by-inch appropriation[2] of the East Room, formerly the "school-room," where she retreats to read and rest:

> The room had become useless, and for some time was quite deserted, except by Fanny… but gradually, as her value for the comforts of it increased, she had added to her possessions, and spent more of her time there; and having nothing to oppose her, had so naturally and so artlessly worked herself into it, that it was now generally admitted to be her's. (508)

Fanny acquires and occupies her sitting room in the same way that a botanical creeper might "naturally and artlessly [work] into" the cran-

[2] Mary Chan also notes Fanny's "slow taking-over of the East room."

nies of a garden wall: "gradually," quietly, and even seasonally. "Mrs. Norris having stipulated for there never being a fire in [the East Room] on Fanny's account," the space's habitability is dictated by the weather, but where territory is concerned, Fanny proves surprisingly tenacious, a hardy vine: "The aspect was so favourable, that even without a fire it was habitable in many an early spring, and late autumn morning, to such a willing mind as Fanny's, and while there was a gleam of sunshine, she hoped not to be driven from it entirely, even when winter came" (508).

I'm not the first to notice Fanny's floral features, and a few critics have recently cast Fanny's coming-of-age story as unusually indebted to late eighteenth- and early nineteenth-century discussions of plants and horticulture.[3] In "*Mansfield Park* as Greenhouse," for example, Mary Chan writes that "Fanny is a science experiment – in a way. Though never intended as such, she is the prize plant in Mansfield Park, my titular metaphorical greenhouse." Chan's primary interest is "the effect of education," and in *Mansfield Park,* she argues, "[d]isposition [...] is formed early and, once formed, is difficult to alter." Fanny, who becomes "a delicate hothouse flower," represents the grafting of gentility – "education and manners" – onto "a natural understanding of what is right and wrong," but in Chan's conception, the Park's other young inhabitants are floral, too. In other words, Chan applies her "titular metaphor" broadly; Fanny may be the "prize plant," but the Bertram children, she intimates, were raised with different degrees of success in the *same* greenhouse: "Fanny is a transplant, one who is bred up *with* her cousins" (my emphasis). When Deidre Shauna Lynch examines the relationship between Romantic-era horticultural practice and Austen's portrayal of female sexuality, she likewise identifies similarities between Fanny's experience and her cousins'. For Lynch, too, the greenhouse looms large in *Mansfield Park*; the precocious Miss Bertrams, she writes, are "like flowers [grown] in enriched soils or in hothouses, perhaps, forced into an over-early bloom" (713), and as the novel progresses, Lynch asserts, the authorities at Mansfield attempt to foster Fanny's growth by applying those same horticultural techniques to her. Working, through a

[3] Akiko Takei's discussion of Fanny and chlorosis, or "green sickness," is also an important part of this conversation. Chlorosis would today be described as iron-deficiency, and Takei's comments on Fanny's greenness resonate with my observations, above, regarding her meat-free diet.

variety of means, to induce his niece to accept Henry Crawford – that is, to bloom sexually in response to Crawford's warm advances – Sir Thomas treats Fanny as he has treated his natural daughters, Mansfield's native cultivars. "[T]he temporal norms that Nature's authority ought to underwrite – concepts of the seasonable, concepts of a proper developmental sequence – are all slightly awry here," Lynch explains, and when Sir Thomas "orders a fire to be laid in the school room," he confirms that "the house of Bertram is a forcing house" (718), a hothouse-esque environment meant to stimulate artificial maturation in *all* the young flowers there rooted, transplants included.

In sum, Chan and Lynch describe Mansfield as a place defined by an association with plants that Austen develops metaphorically and, to some degree, practically; for Chan, the whole place is *like* a greenhouse, and Lynch's Sir Thomas, head gardener, literally forces his charges to bloom in it. I hope, however, to redirect this recent discourse by demonstrating that Austen aligns Fanny – and *only* Fanny – more closely with vegetal organisms than bestial ones. In other words, Chan's and Lynch's botany-focused accounts of Austen's artistry and objectives neglect the novel's careful *comparison* of the vegetal and the bestial, as well as Fanny's unique affiliation with the former category. In Austen's early description of the "well-grown and forward" Miss Bertrams, Lynch sees evidence of hothouse cultivation, of petals unfurled unnaturally early, but Austen's terms arguably speak to the girls' innate animality, to a muscular aggression at odds with Fanny's more plant-like passivity. Maria and Julia pounce, tiger-like, on Fanny's ignorance of English monarchical history – "I cannot remember," one scoffs, "the time when I did not know a great deal that she has not the least notion of yet. How long ago it is aunt, since we used to repeat the chronological order of the kings of England, with the dates of their accession, and most of the principal events of their reigns" (433) – but like violets and tigers, Fanny and her cousins seem, the more closely one looks, like the affiliates of *different* kingdoms, at least from a Linnaean taxonomical perspective. Lynch links the Miss Bertrams' taste for "making artificial flowers" to their status *as* artificial flowers, but since forcing houses are run by the humans of Animalia, not the rosebushes of Plantae (or, to use Linnaeus's term, Vegetabilia), that perspective seems troublingly inconsistent. And indeed, the Bertram girls' precocious sexuality manifests in more than rosy blushes. With regard to men, Fanny's cousins are aggressively territorial; Julia is "fully aware" that Maria's engagement to Mr. Rushworth renders Henry Crawford her "property" (448), and although Maria has, as Austen puts it, "two strings to her bow," her "Rushworth-feelings" subdue her "Crawford-feelings" when she approaches "the vicinity of Sotherton," the *literal* territory that, upon her marriage to Mr. Rushworth, will become her domain (469). Unlike Fanny, who creeps passively into rooms and, later, into hearts, Maria and Julia fight like animals to attain and *re*tain alpha status. "[G]ood friends while their interests were the same," the sisters "had not affection or principle enough to make them merciful or just, to give them honour or compassion" (514), and as the novel progresses, their sexual behavior becomes as animalistic – that is, as unregulated and amoral – as their social behavior. Flirtation and "feelings" give way to illicit touching (most notably during Maria's rehearsals of *Lover's Vows*) and, finally, two elopements. Unmarried Julia absconds with Mr. Yeats, but Maria elopes with Mr. Crawford *after* her marriage to Mr. Rushworth, thus betraying an especially bestial lack of "affection or principle."

If anything, then, Austen's reference to the Miss Bertrams' interest in artificial flowers demonstrates that they have more in common with their forceful father, as Lynch portrays him, than with their creeper of a cousin. However, I also want to challenge the notion that Sir Thomas is first and foremost a horticulturalist. When considered alongside such commonplace terms as "cornfield" or "hayfield," the name "Mansfield" evokes the dystopian image of a plot where men, the most valuable of beasts, are cultivated for consumption, and although Sir Thomas's ostensible Antiguan business is plantation, the troublesome "affairs" (441) that send him abroad are, as Edward Said and his numerous inheritors (including Patricia Rozema, director of the 1999 film adaptation of *Mansfield Park*) have suggested, most likely related to that region's bustling slave trade and, of course, his own slave holdings. In other words, Sir Thomas, master of Mansfield, deals in flesh, not flowers; less ornamental blooms than utilitarian bodies, his daughters, like his slaves, exist to serve the estate. Although Sir Thomas observes upon his return from Antigua that Maria's "behaviour to Mr. Rushworth [is] careless and cold," he is "too glad to be satisfied" with her patently false professions of "happiness" regarding her engagement: "It was an alliance which he could not have relinquished without pain; and thus he reasoned" (534). Despite her animalistic "forward"-ness and acknowledged sentience, Maria's "pain" is trumped by her father's desire to avoid the same; notwithstanding his initial resolution that Maria's "happiness must not be sacrificed to" a convenient but undesired marriage, Sir Thomas recognizes that Mansfield will grow fatter when his daughter is united with a man she "[does] not like," let alone love (534). Not only "happy to escape the embarrassing evils of a rupture," Sir Thomas is "happy to secure a marriage which would bring him *such an addition* of respectability and influence, and very happy to think any thing of his daughter's disposition that was more favourable for the purpose" (534, my emphasis). At this juncture, Austen's repetition of the word "happy" – Sir Thomas is happy, and Maria dishonestly pronounces herself likewise, declaring that she "could not have a doubt of her happiness with [Mr. Rushworth]" (534) – is itself significant, especially in conjunction with Sir Thomas's earlier reference to *sacrifice*, a term that conjures images of animals (often lambs) on altars. Stuart identifies the "happy lamb" as the eighteenth century's "counter-vegetarian mascot," since Alexander Pope and others influentially "[argued] that nature had deliberately kept the lamb devoid of foresight and therefore ignorant of its death until the fatal blow is struck." Citing Pope's "Essay on Man," in which a lamb, "[p]leased to the last," "licks the hand just raised to shed his blood," Stuart articulates one eighteenth-century view of animal slaughter: "The lamb is alive one moment and dead the next: the transition from one state to the other occurs without even a twinge of suffering." "Without pain," he continues, "animal slaughter no longer needed to excite compassion" (218-219). Maria resembles Pope's happy lamb, speaking "securely of her happiness" to her father, who, in turn, proves "very happy to think any thing of his daughter's disposition," so long as it suits his "purpose" (534). Unwilling "to urge the matter quite so far as his judgment might have dictated to others," Sir Thomas resorts to the arguments of a Popean meat-eater, rationalizing his decision with unfounded – indeed, unprovable – assertions about Maria's experience, both present and future: "Her feelings were probably not acute; he had never supposed them to be so; but her comforts might not be less on that account, and if she could dispense with seeing her husband a leading, shining character, there would certainly be every thing else in her favour" (534). In the end, "well-grown" Maria succumbs to the hunger that surrounds her, heightened as it is by her father's "straitened" "means" and "poor returns" (534, 440). As her home's very name implies, the culture at

Mansfield is carnivorous – less greenhouse than slaughterhouse, I might suggest – and Maria differs from the grass-fed lamb only in her taste for meat, in her own willingness to satisfy her desires by consuming her fellows. In her treatment of "nature, animals, and gender" in *Mansfield Park*, Barbara Seeber asserts that "[m]eat is consistently associated with men in the novel" (276), but the Bertram girls represent a sound challenge to the notion that Austen links vegetarian empathy to gender. "Maria felt her triumph, and pursued [Mr. Crawford] careless of Julia" (514), just as Sir Thomas struggles only briefly to reconcile his paternal sentiments with his lordly appetites.

Sir Thomas and his bestial daughters typify the culture of Mansfield, where the subordination of sentience to profit and luxury is standard. The estate's warm-blooded residents and dependents hunt for sport, ride for exercise, consume animal foods with relish – and treat one another with a carnivorous disregard for animal feeling. Lady Bertram, remarkably lazy and presumably corpulent, "[spends] her days sitting nicely dressed on a sofa, doing some long piece of needle work, of little use and no beauty, thinking more of her pug than her children" (434). Notwithstanding her apparent affection for the dog – "I hope she will not teize my poor pug," she frets in anticipation of Fanny's arrival at Mansfield (429) – Lady Bertram more or less discounts the discomfort of any but herself. Indeed, Stuart observes, Austen's culture was like our own: "[T]here was nothing unusual about making elaborate displays of sympathy for animals – fawning over rabbits, lambs or pigeons – and in the next moment devouring the same species of animal at table." In fact, Stuart continues, "[a]s pets – from mollycoddled poodles to caged canaries – became part of popular culture… so farm animals became industrialised as the demand for them burgeoned with affluence. Cruelty exercised in the production of food continued and even increased" (225). The unnamed pug emblematizes the "[c]ruelty" affiliated with both "mollycoddl[ing]" and industrialization, since Lady Bertram eventually pledges her stifled pet's offspring to Fanny, should she consent to marry Mr. Crawford: "And I will tell you what, Fanny – which is more than I did for Maria – the next time pug has a litter you shall have a puppy" (608). In other words, Lady Bertram shares her husband's belief that living bodies, no matter the species, are exchangeable and consumable, and although her willingness to distribute pug's puppies is, perhaps, unremarkable, her abuse of Fanny, selected from her own mother's litter-like "superfluity of children" (426) to be sent to Mansfield, is striking. When Edmund discovers that Fanny's intense headache was triggered by a stint spent cutting roses in the sun on his mother's orders, the "candid Lady Bertram" acknowledges, with astonishing equanimity, her exploitation of her niece: "Poor thing! *She* found it hot enough, but they were so full blown, that one could not wait. [...] I am very much afraid she caught the head-ache there, for the heat was enough to kill any body. It was as much as I could bear myself. Sitting and calling to Pug, and trying to keep him from the flower-beds, was almost too much for me" (464-465). Even the threat of fatality fails to mitigate Lady Bertram's desires, and her sister Mrs. Norris is similarly indifferent to Fanny's pain. If Lady Bertram's "candid" remarks reflect a relatively unvarnished version of the carnivore's creed – in the throes of hunger, "one [cannot] wait" – Mrs. Norris's response to the same headache reflects the perspective of counter-vegetarian thinkers who identified a "mutualistic symbiosis between humans and domesticated animals." Stuart summarizes their position thus: "Using animals for food did not injure them; it actually made them happier: they were bred in prodigious numbers, looked after by assiduous farmers, and then painlessly despatched [*sic*] shortly before becoming too old to enjoy their life in any case" (220). Having insisted that the ailing Fanny "[step] down to [her] house" after cutting roses for Lady Bertram (464), Mrs. Norris asserts, *à la* eighteenth-century counter-vegetarians, that the task "did not injure" Fanny but was designed, rather, to make her "happier." "If Fanny would be more regular at exercise," she says to Edmund, "she would not be knocked up so soon. [...] I thought it would rather do her good after being stooping among the roses; for there is nothing so refreshing as a walk after a fatigue of that kind" (465). To be sure, Mrs. Norris maintains throughout the novel that Fanny, one of a "prodigious" brood, is blessed to be "looked after" as "assiduousl[y]" as she is. Her initial speeches to her homesick young niece en route to Mansfield emphasize Fanny's "wonderful good fortune, and the extraordinary degree of gratitude and good behaviour which it ought to produce" (430), and when, many years later, Fanny later resists taking a role in her cousins' ill-fated production of *Lover's Vows*, Mrs. Norris is quick to rebuke her: "'I am not going to urge her,' – replied Mrs. Norris sharply, 'but I shall think her a very obstinate, ungrateful girl, if she does not do what her Aunt, and Cousins wish her – very ungrateful indeed, considering who and what she is'" (506). From her aunt's perspective, Fanny is a utilitarian creature, more "what" than "who"; in other words, Mrs. Norris, who "spung[es]" cream cheese and pheasant eggs from Sotherton and remarks, with envy sold as abstemiousness,[4] on "the quantity of butter and eggs that were regularly consumed" at the Grants' (483, 440), is a carnivorous spirit. The Grants themselves are kinder to Fanny but less equivocal about their appetites. "[Dr. Grant] was very fond of eating" (440), and he functions, in a way, as an amalgamated caricature of Fanny's two "unreasonable aunts" (465). As lazy as Lady Bertram *and*, somehow, as officious as Mrs. Norris, Dr. Grant is primarily characterized by his taste for flesh. Mary Crawford puts it best when she deems him "an indolent selfish Bon vivant, who must have his palate consulted in everything, who will not stir a finger for the convenience of any one, and who, moreover, if the cook makes a blunder, is out of humour with his excellent wife." "To own the truth," Mary continues, "Henry and I were partly driven out this very evening, by a disappointment about a green goose, which he could not get the better of" (486). Later, Mary reports that Dr. Grant is "ill" as the result of an imperfectly cooked pheasant: "He fancied it tough – sent away his plate – and has been suffering ever since" (519).

Mansfield's younger inhabitants are, typically, as carnivorous as their elders. Tom Bertram, heir to the estate and an avid hunter, reveals his priorities when he sends tidings of his return to Mansfield "first in a letter to the game-keeper, and then in a letter to Edmund" (487).[5] Indeed, Edmund is his primary human victim, since Tom's "extravagance had… been so great as to render" it necessary for his father to bestow upon Dr. Grant the family-living originally intended for Mansfield's second son (436). As Sir Thomas intones "in his most

[4] Seeber identifies other instances in which Mrs. Norris "thinly disguises" her predatory behavior. In addition to "claiming [the cheese and eggs were] all 'quite force[d] upon' her," she offers "unconvincing protestations" regarding her "'feel[ing] for' animals when she attempts to evade Sir Thomas's questions about the theatricals by rapidly changing the topic": "And when we got to the bottom of Sandcroft Hil' […] I could not bear to sit at my ease, and be dragged up at the expense of those noble animals" (272-273).

[5] For Tom, to "respect" animals is to preserve a population large enough to shoot. When his father returns to Mansfield, Tom curries favor with this account of his recent hunting: "[W]e brought home six brace between us, and might each have killed six times as many; but we respect your pheasants, Sir, I assure you, as much as you could desire. I do not think you will find your woods by any means worse stocked than they were" (523).

dignified manner," spendthrift Tom has "robbed Edmund for ten, twenty, thirty years, perhaps for life, of more than half the income which ought to be his," but Tom merely reflects with "cheerful selfishness" that Dr. Grant "would, in all probability die very soon": "[H]e was a short-neck'd apoplectic sort of fellow, and, plied well with good things, would soon pop off" (436). Tom's unceremonious "calculations" (446) regarding the value and duration of his neighbor's life are echoed in Mary Crawford's cool account of her wealthy brother, a confirmed lady-killer if not a would-be parson-popper. "I have three very particular friends," Mary says, "who have been all dying for him in their turn" (447), and although her reference to death-by-love adheres to the figurative conventions of romantic talk, Henry Crawford, having injured Julia and slain Maria, expresses his hunger for Fanny in more credibly predatory terms. When the "forward" Bertram sisters are away from Mansfield, Henry determines "not [...] to eat the bread of idleness," not to go vegetarian – Austen's appropriation of that biblical metaphor seems especially meaningful here – in the face of limited meat (550). Rather, "on the days that [he does] not hunt," Fanny will be his quarry: "I cannot be satisfied without Fanny Price, without making a small hole in Fanny Price's heart. [...] Her looks say, 'I will not like you, I am determined not to like you,' and I say, she shall" (550-551). Like Tom Bertram, Mr. Crawford is "thoughtless and selfish from prosperity and bad example" (488), but whereas Tom's bestial behavior derives, like his mother's, from an outsized concern for his own convenience, Henry declares himself willing to brook any manner of *in*convenience in his pursuit of Fanny's counter-"determined" heart, which he imagines piercing like flesh, not breaking (as convention would dictate) like china. No *idle* sadist, Mr. Crawford takes pleasure in the prospect of a difficult chase – "I never was so long in company with a girl in my life [...] and succeed so ill!" (550-551) – and although his sister pleads briefly for Fanny's sentience – "I will not have you plunge her deep, for she is as good a little creature as ever lived and has a great deal of feeling," she says – Mary ultimately "[leaves] Fanny to her fate" (551). As devoted to the consumption of women as Dr. Grant, his gluttonous host, is fanatical about the consumption of poultry, Henry means to devour Fanny whole, and Mary, "without attempting any farther remonstrance," jokes about the "[m]oderation" of his proclamation. "No, I will not do her any harm, dear little soul!" Henry says. "I only want her to look kindly on me, to give me smiles as well as blushes, to keep a chair for me by herself wherever we are, and be all animation when I take it and talk to her; to think as I think, be interested in all my possessions and pleasures, try to keep me longer at Mansfield, and feel when I go away that she shall be never happy again. I want nothing more" (551). The intensity of Mr. Crawford's solipsism is undeniable, and when the prospect of Fanny's death again enters the discourse, he reasons tautologically – or, we might say, carnivorously – that if his frivolous suit kills Miss Price, she was destined to die anyway, to have been unable "to enjoy her life in any case" (Stuart 220): "'It can be but for a fortnight,' said Henry, 'and if a fortnight can kill her, she must have had a constitution which nothing could save'" (551).

When Mr. Crawford takes explicit aim at Fanny, Austen's narrator joins the conversation with an "I" of her own:

> [A]lthough there doubtless are such unconquerable young ladies of eighteen (or one should not read about them) as are never to be persuaded into love against their judgment by all that talent, manner, attention, and flattery can do, I have no inclination to believe Fanny one of them, or to think that with so much tenderness of disposition, and so much taste as belonged to her, she could have escaped heart-whole from the courtship (though the courtship only a fortnight) of such a man as Crawford, in spite of there being some previous ill-opinion of him to be overcome, had not her affection been engaged elsewhere. (551)

This intervention is significant. Despite what I've described as her unique affiliation with plants and, even at times, her relative similarity to vegetable matter, Fanny *is* a person, a woman of "taste" and "tenderness," of aching head and muscular heart, who differs from other "young ladies of eighteen" – that is, from the many girls "dying for" Henry Crawford – in one factor only: "[H]er affection," Austen specifies, is "engaged elsewhere." At the most literal level, "elsewhere" is, of course, with Edmund, the figure in the novel who most consistently defends Fanny's human agency while simultaneously encouraging her Romantic veneration of the natural (as opposed to the cultured) world. As I suggested above, Fanny's vine-like attachment to Mansfield is initially fostered when Edmund "encouraged her taste, and corrected her judgment," and she wilts like Shelley's Sensitive Plant when, in the absence of her curatorial cousin, she's conscripted to cut roses in an environment that's hostile to her physically and emotionally sensitive constitution. More metaphorically, however, the "elsewhere" on which Fanny's affection is fixed is an abstract moral realm in which humane consideration, that cornerstone of vegetarian belief, is an operative value – a realm that Edmund, as Fanny idealizes him, more promises than unfailingly exemplifies. Both a "companionless Sensitive Plant" who relies on Edmund's attentive care *and*, as Austen emphasizes, a woman akin to the Lady of "lovely mind" (II.5) who, before her death, tends Shelley's garden "from morn to even," Fanny resembles both central figures – one vegetal, one human – in Shelley's eccentric meditation on life and feeling. In other words, creeperish Fanny is herself a sensitive planter, a gardener of sorts. She goes to the East Room to "[visit] her plants,"[6] and her willingness to acknowledge but forgive "the pains of tyranny, of ridicule, and neglect" (508) inflicted by her cousins in their former school-room resembles the readiness of Shelley's Eve-like protectress to relocate, tenderly and humbly, her garden's malignant animals:

> And all killing insects and gnawing worms,
> And things of obscene and unlovely forms,
> She bore, in a basket of Indian woof,
> Into the rough woods far aloof,
>
> In a basket, of grasses and wild-flowers full,
> The freshest her gentle hands could pull
> For the poor banished insects, whose intent,
> Although they did ill, was innocent. (II.41-48)

Shelley's account of "[t]his fairest creature" (II.57) is undeniably vegetarian in its emphasis on the respect that she rightfully accords every living creature, "unlovely" or otherwise, in that "undefiled Paradise" (I.58), and as *Mansfield Park* progresses, Fanny becomes an increasingly vocal proponent of similar values. Although Fanny retains her *vegetal* features, she also embraces, with the slow surety of a climbing vine, a new *vegetarian* identity. Can a "domestic sphere" experience revitalization when its human inhabitants learn to resist, in deed and speech, the kind of consumption that "animalises" in favor of the sort

---

[6] Deresiewicz observes that "the East room, with its houseplants … makes a gesture toward the verdant—a silly idea, this would be, were these not the only houseplants in all of Austen, appearing just here, of all places" (58).

that sensitizes? That Shelleyan question is at the heart of Austen's novel, and since Henry Crawford's contrived attachment to Fanny develops late in the narrative and without his knowing anything of Fanny's long-standing "affection" for Edmund, it is a question that Austen explores primarily through the amorous triangle that includes Mary, Edmund, and Fanny. Edmund initially reciprocates Mary's admiration despite his awareness of her moral shortcomings, but Fanny, wholly committed to a Cheynian-Shelleyan set of ethical principles, comes to exert considerable influence over Mary and Edmund alike when Sir Thomas returns from Antigua and the battle for Mansfield's ethical future intensifies.

A confirmed urbanite, Mary early demonstrates the degree to which she's out of touch with the rhythms and priorities of the agricultural world when she offends the neighborhood farmers – "all the farmers, all the labourers, all the hay in the parish," as she glibly puts it – by attempting to hire a horse and cart to convey her harp to the parsonage in the midst of what Edmund terms "a very late hay harvest" (456). "Guess my surprise, when I found that I had been asking the most unreasonable, most impossible thing in the world," Mary declares. "I shall understand your ways in time; but coming down with the true London maxim, that every thing is to be got with money, I was a little embarrassed at first by the sturdy independence of your country customs" (456). In fact, Mary's misuse of horses and denigration of planting – "I would have every thing as complete as possible in the country, shrubberies and flower gardens [...] but it must all be done without my care," she says immediately before giving "tidings of [her] harp" (455) – only intensify in subsequent pages. First, Miss Crawford expresses "an inclination to learn to ride" (460). Edmund's "quiet mare," which had been designated for Fanny's use, is loaned "for the purpose," and although Edmund intends for Fanny to sustain "[n]o pain, no injury" – "[Fanny] was not to lose a day's exercise" as a result of Mary's lessons – Mary's second day of riding with Edmund "was not so guiltless" as her first, when "she [had] returned [...] in excellent time": "Miss Crawford's enjoyment of riding was such, that she did not know how to leave off. Active and fearless, and, though rather small, strongly made, she seemed formed for a horse woman [...] [and was] unwilling to dismount" (460-461). Fanny, "ready and waiting" for her turn, "wondered that Edmund should forget her," and she determines, with a vegetarian's aspiration to forestall suffering –

recall Shelley's sympathy for those "poor banished insects" – not to forget the horse: "She began to think it rather hard upon the mare to have such double duty; if she were forgotten, the poor mare should be remembered" (461). Mary, in contrast, expresses no real compunction regarding her careless attitude toward Fanny's time and the mare's labor: "I have nothing in the world to say for myself – I knew I was very late, and that I was behaving extremely ill; and therefore, if you please, you must forgive me. Selfishness must always be forgiven you know, because there is no hope of a cure" (461-462). "Nothing ever fatigues me, but doing what I do not like," she continues, and by "the next day," Mary has, with Edmund's assistance, claimed the mare "for a whole morning" (462). Within the space of half a paragraph, Mary has ridden the mare "[f]our fine mornings successively" (463), and it's her insatiable appetite – a taste of riding leads quickly to a hunger for it – that Austen's narrative emphasizes. Her well-fed host "will not stir a finger for the convenience of any one," but Mary isn't as unlike Dr. Grant as her own critique implies. In both cases, rapacious consumption of animals – Mary consumes the horse's energy, while Dr. Grant, gourmand, consumes bodies more directly – leads to indifferent treatment of humans; when the cook errs, Dr. Grant is "out of humour with his excellent wife," while Mary neglects Fanny, who, as Edmund himself acknowledges, rides less for pleasure than for health (462). Moreover, both Mary and Dr. Grant share, in their carnivorous obsession with animals, a relative disdain for plants. The scuffle between Mrs. Norris and Dr. Grant regarding the quality of the parsonage's apricot tree reveals that Dr. Grant "hardly knows what the natural taste of [the] apricot is": "'The truth is, Ma'am,' said Mrs. Grant [...] 'ours is such a remarkably large, fair sort, that what with early tarts and preserves, my cook contrives to get them all'" (454). The Grants' habit of obliterating "the natural taste" of their fruit is consistent with their carnivorous diet; many vegetarians advocated the consumption of raw (or, at worst, lightly cooked) produce, since, as the seventeenth-century vegetarian mystic Thomas Tryon argued, consuming cooked fruits was almost as likely as consuming flesh to "[block] the circulatory system of the body": "Plantains [for example] need not be cooked in pies or tarts because in their natural state they have already reached a state of sufficient richness [...] '[I]f baked with Sugar, Spices and the like [they] lose their natural operation, and thereby become hot, and apt to obstruct the Passages, and tire the Appetite and Stomach, generate evil Juices, dull Spirits and thick Blood'" (Smith 110-111). Mary is as unimpressed by vegetation, edible or otherwise, as Dr. Grant. En route to Sotherton, riding "through a pleasant country," Fanny observes with pleasure even "the difference of soil," but Mary "saw nature, inanimate nature, with little observation; her attention was all for men and women" (469). When, later in the novel, Fanny "rhapsodiz[es]" about laurels, evergreens, and, astonishingly, soil again, she tellingly invokes an eating metaphor – "One cannot fix one's eyes on the commonest natural production without finding food for a rambling fancy" – but flesh-fed Mary remains unable to share her ardor: "'To say the truth,' replied Miss Crawford, 'I am something like the famous Doge at the court of Lewis XIV; and may declare that I see no wonder in this shrubbery equal to seeing myself in it'" (539).

However, during the same *tête-à-tête* in which Mary declares herself more French sophisticate than English Romantic, she also acknowledges that, "take it all and all, [she] never spent so happy a summer" as the present one, and she goes so far as to state, "I am conscious of being far better reconciled to a country residence than I had ever expected to be" (539). Fanny recognizes that Mary's admission, made with "brightened eyes" (539), is due primarily to her feelings for Edmund, but one can't dismiss the role that Fanny herself has played in Mary's conversion. When the Crawfords first arrive at the parsonage, Mrs. Grant declares that "Mansfield shall cure you both," and although she refers primarily to her half-siblings' negative perception of matrimony – "Every body is taken in at some period or another," shrugs Mary – her medical diction recalls Dr. Cheyne's proclamations regarding the far-reaching effects of vegetarian living. Notwithstanding Mansfield's violent name and Mary's own assertion that "[s]elfishness" has "no hope of a cure," Mary *does* change as a result of her contact with Fanny, who lives, as Cheyne might put it, "strictly [...] and vegetably." Not long after horse-dominating Mary declares herself hopelessly egoistic, she reacts with horror to Mrs. Norris's humiliatingly public assertion that Fanny is "a very obstinate, ungrateful girl," and when she quietly intercedes on Fanny's behalf, she bests even Edmund:

> Edmund was too angry to speak; but Miss Crawford looking for a moment with astonished eyes at Mrs. Norris, and then at Fanny, whose tears were beginning to show themselves, immediately said with some keenness, "I do not like my situa-

> tion; this *place* is too hot for me" – and moved away her chair to the opposite side of the table close to Fanny, saying to her in a kind low whisper as she placed herself, "Never mind, my dear Miss Price... do not let us mind them;" and with pointed attention continued to talk and endeavor to raise her spirits, in spite of being out of spirits herself. (506)

Stung by Edmund's initial refusal to take the role of Anhalt in *Lover's Vows*, Mary had, only moments earlier, "moved her chair considerably nearer the tea-table," where she "[had given] all her attention to Mrs. Norris, who was presiding there" (505). Now, however, Mary associates the carnivorous Mrs. Norris with painful heat, and when she consciously exchanges her excessively warm "place" for one on "the opposite side" of a table laden with consumables, she symbolically acknowledges the brutality inherent in the act of cooking and eating – or, in this case, publicly *roasting* and cruelly using – a sentient creature. When Tom proposes that Fanny act in the play, he disregards her sentience, describing her as a purely utilitarian animal: "[I]t will not much signify if nobody hears a word you say, so you may be as creepmouse as you like, but we must have you to look at" (505). Mrs. Norris confirms Tom's perception of Fanny's value, but Mary, "despite being out of spirits herself," resists it; her behavior, Austen writes, is "almost purely governed" by "really good feelings," and her new vegetarian-style self*less*ness finds favor not only with Edmund but also with Fanny herself. "Fanny did not love Miss Crawford," Austen writes, "but she felt very much obliged to her for her present kindness." As Mary demonstrates all that she has, quite surprisingly, learned about Fanny's priorities during their limited intercourse – in addition to complimenting Fanny on her needlework, she "enquire[s] if [Fanny] had heard lately from her brother at sea" – Fanny "could not help admitting it to be very agreeable flattery, or help listening, and answering with more animation than she had intended" (506). Mary's act not only preserves Fanny's dignity but also, Austen's diction intimates, saves her life. Fanny responds involuntarily to Mary's resuscitating humanity – she "could not help [...] listening, and answering" – which results in greater "animation," and Mary, too, seems encouraged by the success of her compassionate intervention. As Shelley suggests, to tend a Sensitive Plant is to develop one's own capacity for sensitivity – to be cultivated, however counterintuitively, by the cultivar. In order to prevent the unfamiliar Mr. Maddox from playing opposite Miss Crawford, Edmund agrees to take the role of Anhalt – "She has a right to be felt for," Edmund claims, "because she evidently feels for herself [...] Her feelings ought to be respected" (510) – and Mary subsequently visits Fanny in the geranium-bedecked East Room "on purpose to intreat [her] help" with lines (517). The intimacy between Mary and Fanny continues to grow, and although Fanny recognizes that Mary's interest in the relationship is due at least in part to boredom – when Fanny turns up at the parsonage "on a wet day" after the Bertram sisters have left Mansfield, Mary regards her as "an event," an unexpected diversion that renders her "all alive again directly" (537) – the two women nonetheless "[saunter] about together many an half hour in Mrs. Grant's shrubbery" throughout the fall, "venturing sometimes even to sit down on one of the benches now comparatively unsheltered, remaining there perhaps till in the midst of some tender ejaculation of Fanny's, on the sweets of so protracted an autumn, they [are] forced by the sudden swell of a cold gust shaking down the last few yellow leaves about them, to jump up and walk for warmth" (538). In sum, despite her enduring preference for "men and women" over "inanimate nature," Mary's increasingly frequent interaction with Fanny constitutes a kind of vegetable-immersion therapy, and notwithstanding his romantic regard for Miss Crawford, even Edmund does not "consider Fanny as the only, or even as the greater gainer by such a friendship" (540). When Mary wishes to be lovingly scolded for sitting outdoors with Fanny in cold weather – "What do you think we have been sitting down for but to be talked to about it, and entreated and supplicated never to do so again?" – Edmund replies with Cheynian equanimity to the perceived difficulties of living "vegetably": "[R]eally [...] the day is so mild that your sitting down for a few minutes can be hardly thought imprudent. Our weather must not always be judged by the calendar. We may sometimes take greater liberties in November than in May" (540).

The first figure to demonstrate genuine interest in plant-like Fanny and the person from whom Fanny learns to appreciate the natural world, Edmund himself boasts a long-standing respect for the vegetable realm and, unlike his parents and siblings, an innate benevolence. The trouble, though, is that Fanny's passive conversion of Mary Crawford parallels Mary's more active attempts to convert Edmund to her own more bestial values. Fanny affirms her own passivity when, pressed to take part in her cousins' theatricals, she breathlessly declares, "Indeed you must excuse me. I could not act any thing if you were to give me the world. No, indeed, I cannot act" (505). Mary, in contrast, has already agreed to play Amelia – "[a]n odious, little, pert, unnatural, impudent girl," as Julia angrily depicts the character (500) – and despite Edmund's concerns regarding the appropriateness of home theatricals generally and of *Lover's Vows* in particular, he eventually compromises his own moral code and, as I've already noted, accepts the role of Anhalt. Cheyne had warned Richardson that it would be difficult to resist the "free-livers" and "Beef-eaters," and Edmund struggles similarly in the face of Mansfield's "Voluptuous" company. When, to Fanny's great surprise, Edmund seeks her advice regarding his participation in the play, he sees that "[her] judgment is not with [him]," but his ostensible concern for Mary's "feelings" win the day (510). Edmund claims to have no choice – "I can think of no other alternative," he says (509) – but Fanny, having escaped to the East Room to "[give] air to her geraniums" and perhaps "inhale a breeze of mental strength herself" (509), responds to Edmund with a quiet appeal to principles, as opposed to pragmatics. "I am sorry for Mary Crawford," she says, "but I am more sorry to see you drawn in to do what you had resolved against … It will be such a triumph to the others" (510). Fanny's prediction regarding the "triumph" of her greedy cousins proves correct: "Such a victory over Edmund's discretion had been beyond [Tom's and Maria's] hopes, and […] they congratulated each other in private on the jealous weakness to which they attributed the change, with all the glee of feelings gratified in every way. Edmund […] was to act, and he was driven to it by the force of selfish inclinations only. Edmund had descended from that moral elevation which he had maintained before, and they were both as much the happier for the descent" (511). Neither the whiff of geraniums nor the wise words of the cousin he's trained like ivy can fortify Edmund in the face of Mary's carnal power; Miss Crawford later remarks that she experienced "exquisite happiness" when she watched Edmund's "sturdy spirit […] bend as it did": "Oh! it was sweet beyond expression" (622). Edmund's hypocritical behavior—he insists, unconvincingly, that it's just "the *appearance* of inconsistency" (509) – resonates with the vegetarian ideology expressed in Richardson's *Clarissa*. "When Lovelace's hirelings' sympathies are finally reawakened they still receive pecuniary benefits from Clarissa's plight and do nothing to save her," Stuart notes. "This, Richardson indicates, is the equivalent of claiming to sympathise [with] animals but continuing to eat them. 'A pitiful fellow!' exclaims Lovelace of one of them. 'Such a ridiculous kind of pity *his*, as those silly souls have, who would not kill an innocent chicken for the world; but when killed to their hands, are always the most greedy devourers of it'" (191). Indeed, Edmund increasingly resembles those "hirelings" when, in the context of Henry Crawford's Lovelace-esque determination to penetrate Fanny's unwilling heart, he abandons the nurturing role that, with regard to his shy cousin, he has always played. Despite hours spent absorbing Fanny's "tender ejaculation[s]" concerning the beauty of the shrubbery through which they tramped together, Mary only minimally protests her rakish brother's plan to annihilate Fanny's happiness, and although Edmund himself is unaware that Henry's suit began as frivolously as it did, he certainly stands to gain – practically, if not monetarily – from its success. Fanny's union with Henry would facilitate Edmund's own success with Mary, Henry's beloved sister, and although Edmund initially praises Fanny for refusing to concede to "marriage without love" (615), he warmly encourages her to sacrifice her emotional agency: "But [...] let him succeed at last, Fanny, let him succeed at last. You have proved yourself upright and disinterested, prove yourself grateful and tenderhearted; and then you will be the perfect model of a woman, which I have always believed you born for" (616). In the light of Henry's original desire to leave "a small hole" in Fanny's heart, Edmund's use of the word "tenderhearted" renders him complicit, albeit after the fact, in Henry's carnivorous pursuit; Mr. Grant once "fancied" his pheasant to be "tough," but Fanny, Edmund asserts, should "prove" herself tender, fork-friendly. Having compromised his principles for Mary's sake in the case of the play, Edmund's treatment of Fanny, whose sentience and judgment he once championed, more and more resembles that of Lady Bertram and Mrs. Norris, who value only "her habits of ready submission" (621).

And yet, Mary's attempts to "animalise" – I borrow Shelley's term – Edmund's attitudes encounter surprisingly potent resistance. Upon finding "a nice little wood" at Sotherton barred by a locked gate, Mary had earlier remarked with frustration that "in these great places, the gardeners are the only people who can go where they like" (474). Her comment is prescient; a hybrid of plant and planter, Fanny comes to wield a power at Mansfield that denies the logic of class and culture. Mary herself remains unsuspectingly affected by Fanny's passive "cure," and Deresiewicz is rightly struck by the degree to which Mary, despite her willingness to abandon Fanny to Henry's selfish "courtship," actually *resembles* Fanny during her own final visit to the East Room, when she slips into "a reverie of sweet remembrances" (623). Some pages earlier, Romantic Fanny had remarked on the "very wonderful [...] operations of time, and the changes of the human mind" – "If any one faculty of our nature may be called *more* wonderful than the rest, I do think it is memory," she says – but Mary, "untouched and inattentive, had nothing to say" (538-539). In her own later "remembrances," however, Mary – "one of those unfixed, heartless, amnesiac Crawfords," as Deresiewicz describes her – seems to have "discovered what it means to have a heart" (Deresiewicz 83), and that development foreshadows Fanny's ultimate success with regard to Edmund and, more broadly, Mansfield Park. The same tenacity that allows Fanny to occupy the chilly East Room while any "gleam of sunshine" yet remains allows her to cling, creeper-like, to her principles despite the emotional adversity that surrounds her. When, at the end of Fanny's coming-out ball, Sir Thomas advises – "'Advise,'" Austen notes, "was his word, but it was the advice of absolute power" – his niece "to go immediately to bed," he means "to recommend her as a wife by shewing her persuadableness." Fanny obligingly "creep[s] slowly up the principal staircase" (579), but Sir Thomas fails to recognize that "creeping slowly," far from indicating Fanny's "persuadableness," emblematizes the ivyish strength through which she has acquired the East Room and, following Mr. Crawford's proposal, with which she pursues her pseudo-vegetarian agenda. The *Lover's Vows* debacle has already demonstrated that Fanny's "habits of ready submission" have their limits, and as the novel rolls to its close, Fanny proves once and for all that to live with and *as* plants, not animals, is the surest path to health and happiness.

It's no coincidence that the woman who reveres vegetation – "Cut down an avenue! What a pity!" she gasps when Mr. Rushworth reveals his plans for Sotherton's ancient trees – and "remember[s]" the feelings of an overworked mare is the only figure in *Mansfield Park* to acknowledge that the Bertrams' wealth (and Sir Thomas's Antiguan difficulties) are born of slavery, an institution that, figuratively speaking, qualifies as both carnivorous and cannibalistic.[7] "Did not you hear

[7] Seeber thus describes the relationship between slavery and consumption: "The instrumental view of nature and animals provides a legitimating master-paradigm for

me ask him about the slave trade last night?" Fanny inquires when Edmund encourages her to speak more frequently to her uncle "in the evening circle" (532). "I longed [to ask additional questions]," Fanny continues, "but there was such a dead silence! And while my cousins were sitting by without speaking a word, or seeming at all interested in the subject, I did not like ... to set myself off at their expense" (533). However, although Fanny hesitates to discuss slavery abroad, she proves less apprehensive about resisting it at home. Sir Thomas jumps at the chance, afforded by Mr. Crawford's offer of marriage, to use Fanny as he had used Maria – that is, to exchange her hand for "fortune" and social prestige (597) – and he is both "disappointed" and angry to discover that Fanny is *not* "free from wilfulness of temper [...] and independence of spirit" (599). When Fanny makes her refusal clear, Sir Thomas declares himself "half inclined to think [that she] does not quite know [her] own feelings" (598), but when he insists that Fanny take a composing turn outdoors – "I advise you to go out, the air will do you good; go out for an hour on the gravel, you will have the shrubbery to yourself, and will be the better for air and exercise" – he demonstrates his profound ignorance of her character. Sir Thomas imagines that a quiet walk will help Fanny "to reason [herself] into a stronger frame of mind," and although his vision is, to some degree, correct, his "better" is not consistent with his niece's. The young woman who, in the East Room, "inhale[s] [...] mental strength" from the same breeze that fortifies her geraniums "joyfully [obeys]" her uncle's instructions, but her jaunt through Mansfield's vegetation only strengthens her belief in her *own* convictions (601). Notwithstanding Sir Thomas's violent disapprobation of her refusal of Henry, Fanny feels, following her restorative encounter with the shrubbery, that "the evening set in with more composure [...] and more cheerfulness of spirits than she could have hoped for": "[S]he trusted [...] that she had done right, that her judgment had not misled her" (602).

In subsequent pages, as carnivorous forces assail her from every direction, Fanny articulates her vegetarian values with increasing clarity and specificity. Although Mr. Crawford was "very much in love," it was "a love which, operating on an active, sanguine spirit," possessed "more warmth than delicacy," and Fanny, "sanguine" in neither

---

the systematic domination of women and the exploitation of Antigua (on which the wealth of the Bertrams depends)" (273).

tastes nor affect, is angered by Henry's plan "[to force] her to love him" (603). "Now she was angry," Austen writes. "Some resentment did arise at a perseverance so selfish and ungenerous. Here was again a want of delicacy and regard for others which had formerly so struck and disgusted her [...] How evidently was there a gross want of feeling and humanity where his own pleasure was concerned" (605). In her careful juxtaposition of "others[']" feelings and one's "own pleasure," Fanny recapitulates Cheyne's argument regarding sympathy and diet. For Cheyne, Stuart explains, "'[s]ympathy' was a natural function of the nervous system which caused people to experience 'the Misery of their Fellow Creatures' and this was the physiological basis of social harmony" (176). Henry refuses to recognize (let alone "experience") Fanny's "Misery" despite her *own* attempts to sympathize with *him* – "[H]e approached her now with rights that demanded different treatment. She must be courteous, and she must be compassionate" (604) – and the single-mindedness with which he pursues "his own pleasure" affiliates Mr. Crawford, in Fanny's perception, with Cheyne's morally perverted "Flesh-eater[s]." Fanny vigorously defends her *unique* right to considerate treatment – "How was I... to be in love with him the moment he said he was with me? How was I to have an attachment at his service, as soon as it was asked for? His sisters should consider me as well as him," Fanny says to Edmund (619) – but even more importantly, she *generalizes* about sympathy and human experience. In short, she begins to formulate a treatise, a Shelleyan *Vindication*, of her own. "Let him have all the perfections in the world," Fanny asserts with aphoristic aplomb, "I think it ought not be set down as certain, that a man must be acceptable to every woman he may happen to like himself" (619), and when Mary encourages her to rejoice in "the glory of fixing one who has been shot at by so many," Fanny resists being cast as a huntress, choosing instead to defend Henry's (and others') sentient hunted: "I cannot think well of a man who sports with any woman's feelings; and there may often be a great deal more suffered than a stander-by can judge of" (624-625). Mary, of course, is well aware of the suffering that derives from such sport. Her good friend Flora Ross was, as Mary tells it, "dying for Henry the first winter she came out," but, threatened by the consume-or-be-consumed "jungle" that Deresiewicz describes, Flora "jilted a very nice young man [...] for the sake of that horrid Lord Stornaway," who, according to Miss Crawford, "has about as much sense as [...]

Mr. Rushworth, but [is] much worse looking, and with a blackguard character" (624). Flora, Austen intimates, would have done better to have lived up to her name, to have eschewed the bloody business of carnivorous pursuit in favor of sympathetic, vegetarian-style living.

Despite Mary's verbal manipulation and Sir Thomas's decision to send her to Portsmouth, where, he reasons, she will regret "a little abstinence from the elegancies and luxuries of Mansfield Park" (628), Fanny, whose bloom was never conventional – "Pray, is she out, or is she not?" Miss Crawford asks early in the novel, before stating that "[t]ill now, [she] could not have supposed it possible to be mistaken as to a girl's being out or not" (450) – proves a less corruptible flower than Flora. Sir Thomas imagines that the visit to Portsmouth will function as "a medicinal project on his niece's understanding, which he must consider as at present diseased" (628), but Fanny, the original cure, resists her uncle's prescription. Although her lengthy stay at her parents' urban home compromises Fanny's vigor – "[S]he had lost ground as to health since her being in Portsmouth [...] [H]er face was less blooming than it ought to be" (651) – she remains, unlike the novel's "amnesiac" others, committed to living "strictly [...] and vegetably." When Edmund encounters Miss Crawford in London, he finds her "altered" for the worse – "It is the influence of the fashionable world [...] It is the habits of wealth," he writes despondently to Fanny (657) – and when Mary speaks "[s]o voluntarily, so freely, so coolly" of Maria's adulterous elopement with Henry, Edmund's "eyes are opened" to her "corrupted, vitiated mind" (675, 676). He only learns later that she had harbored a greedy wish for his brother's death. "I really am quite agitated on the subject," Mary had written to Fanny on the occasion of Tom's illness. "Fanny, Fanny, I see you smile, and look cunning, but upon my honour, I never bribed a physician in my life" (664). Once nearly "cured" of debilitating egoism by a Cheynian "diet" of Fanny's company, Mary not only regresses to a state of bestial insensitivity but perversely defines it as the opposite: "And now, do not trouble yourself to be ashamed of either my feelings or your own. Believe me, they are not only natural, they are philanthropic and virtuous. I put it to your conscience, whether 'Sir Edmund' would not do more good with all the Bertram property, than any other possible 'Sir'" (664).

As her references to Humphry Repton and "improvement" suggest, Austen wrote with a keen awareness of the trends – political, aesthetic, and philosophical – that defined the world around her, but in *Mansfield Park,* Repton-style modernization fails as a means of securing felicity, conjugal or otherwise. Meanwhile, Mansfield's homesteads are physically unchanged but nonetheless much improved by the novel's conclusion. Shifts in perspective – Deresiewicz might call them "substitutions" – play a role in those improvements, or perceived improvements. Fanny, whose nostalgia for Portsmouth had once been acute, comes to realize after a months-long stay with her parents that "Mansfield was home" (662), and she lovingly recalls her uncle's estate as a place where "every body had their due importance; [where] every body's feelings were consulted" (641). Fanny herself, of course, had never been accorded much "importance" or respect at Mansfield, but her generous appraisal of her adoptive home isn't mere delusion. During her years at Mansfield, Fanny had sown the seeds of consideration, and although they had only begun to sprout (as evidenced, for instance, by Mary's partial metamorphosis) at the time of her withdrawal to Portsmouth, she maintains a gardener's confidence in the arrival of fruit, as well as a vegetarian's confidence in that fruit's capacity to engender radical transformation. As Fanny comes to recognize when she faces the prospect of spending the entire spring in the gloomy environs of a city, natural growth is slow but steady:

> She had not known before, how much the beginnings and progress of vegetation had delighted her. – What animation both of body and mind, she had derived from watching the advance of that season which cannot, in spite of its capriciousness, be unlovely, and seeing its increasing beauties, from the earliest flowers, in the warmest divisions of her aunt's garden, to the opening of leaves of her uncle's plantations, and the glory of his woods. (663)

If Repton-esque "improvement" falls short in *Mansfield Park,* the "progress of vegetation" decidedly prevails, and when Fanny at last returns to Mansfield, she realizes the degree to which her vegetal spirit and vegetarian ethics have *made* that place the idealized home of her fancy. As Fanny walks through the door, Lady Bertram gives voice to her own relief – "Dear Fanny! now I shall be comfortable!" – but nonetheless leaves "the drawing room to meet her [niece]," walking "with no indolent step" (671). For his part, Sir Thomas is "[s]ick of ambitious and mercenary connections, prizing more and more the sterling good principle of temper" (685), and within paragraphs, Fanny not only "[settles] at Thornton Lacey" (686) as Edmund's wife but becomes mistress of the Mansfield parsonage, the house associated, over the course of the novel, with pheasant-"spunging" Mrs. Norris, horse-abusing Mary Crawford, and, most importantly, goose-obsessed Dr. Grant. As a result, perhaps, of his un-Cheynian habits, Dr. Grant dies not long after Fanny's and Edmund's marriage, but not having "plied [him] [...] with good things" *themselves*, the young couple needn't worry that "the acquisition of [the] Mansfield living [...] occurred just after they had been married long enough to begin to want an increase of income, and feel their distance from the paternal abode an inconvenience" (686). After all, vegetation's subversive progress derives in part from its adaptability, its variety. When spring finally returns to the Sensitive Plant's garden, "the mandrakes, and toadstools, and docks, and darnels,/Rose like the dead from their ruined charnels" (III.110-113), and when Fanny moves to the parsonage, a place she "had never been able to approach but with some painful sensation of restraint or alarm," it "soon grew as dear to her heart, and as thoroughly perfect in her eyes, as every thing else, within the view and patronage of Mansfield Park had long been" (686).

Works Cited

Austen, Jane. *The Complete Novels*. Intro. Karen Joy Fowler. New York: Penguin, 2006. Print.

—. *Selected Letters*. Ed. R.W. Chapman. Intro. Marilyn Butler. Oxford: Oxford UP, 1985. Print.

Chan, Mary M. "Mansfield Park as Greenhouse: 'The Effect of Education' in *Mansfield Park. Persuasions On-Line* 27.1 (2006): no pagination. Print.

Deresiewicz, William. *Jane Austen and the Romantic Poets*. New York: Columbia UP, 2004. Print.

Lynch, Deidre Shauna. "'Young Ladies Are Delicate Plants': Jane Austen and Greenhouse Romanticism. *ELH* 77 (2010): 689-729. Print.

Seeber, Barbara K. "Nature, Animals, and Gender in Jane Austen's *Mansfield Park* and *Emma*." *Literature Interpretation Theory* 13 (2002): 269-285. Print.

Smith, Nigel. "Enthusiasm and Enlightenment: of food, filth, and slavery." *The Country and the City Revisited: England and the*

Politics of Culture, 1550-1850. Eds. Gerald MacLean, Donna Landry, and Joseph Ward. Cambridge: Cambridge UP, 1999. Print.

Shelley, Percy Bysshe. "The Sensitive Plant." *English Romantic Writers*. 2nd ed. Ed. David Perkins. Fort Worth: Harcourt Brace College Publishers, 1995. Print.

Stuart, Tristram. *The Bloodless Revolution: A Cultural History of Vegetarianism from 1600 to Modern Times*. New York: W.W. Norton and Co., 2006. Print.

Takei, Akiko. "'Your Complexion Is So Improved!': A Diagnosis of Fanny Price's 'Dis-ease.'" *Eighteenth-Century Fiction* 17.4 (2005): 684-700. Print.

Plants and the Problem of Authority in the Antebellum U.S. South

Lynne Feeley

Abstract: This essay attends to the language of antebellum farm journals and plantation records to argue that Southern planters encountered resistance from their lands and crops that they could not countenance. Planters' failure to adapt to their environments stemmed from a set of assumptions about their role as masters over the natural world. In this essay, I argue that planters refused to comprehend the messages being sent by their unruly land because they feared conceding any of their supposed authority, for to do so would have been to compromise the "naturalness" of their control over their enslaved people.

In July of 1857, overseer John Webster dispatched the following report to his plantation's absentee owner, Paul Cameron: "We have had some strange wether heare for the last ten days. We have had in that time five frosts too freses and the biggest snow I ever saw sinse I have been in this state."[1] Webster made this astonishing report from Greene County, Alabama, where Cameron, a powerful North Carolina planter, had migrated in search of that so-called "white gold." But Cameron's Greene County plantation failed, and failed spectacularly. Season after season, disease, pests, and rains destroyed his crops; snow in July, when Alabama's thick heat ought to have been nurturing the cotton plants toward their September harvest, was only the most radical display of Greene's unruly and unpredictable climate. Planters' account ing of antebellum agriculture in the farm journals and letters studied here reveal that their plants greatly distressed them. I mean "dis-

[1] All citations regarding Cameron's plantations in Alabama and North Carolina are from the Cameron Family Papers #133, Southern Historical Collection, Wilson Library, University of North Carolina at Chapel Hill. In-text citations from this collection include the writer and recipient of the letter and the date. I have retained all original spelling and punctuation except where errors render the sentence incomprehensible, in which cases I have added punctuation in brackets.

tressed" in two senses. In the first sense, their crops caused them strife, reflected in their affect-laden agricultural discourse. But in the sense that "distress" signifies an overpowering, adverse pressure that compels action or movement, plants, and the profound profits they promised but often seemed to withhold, compelled planters' migration from the Old South to the Deep South, as they searched for land and for crops that would submit to their authority.

This essay traces the language of hope and distress in antebellum agricultural discourse to argue that soils and plants vexed planters, but that planters found it necessary to disavow plants' unruliness. This disavowal stemmed from the imperative for planters to view themselves as unchallenged masters of the earth, an imperative bound together with their self-positioning as rightful masters of non-white peoples. In other words, I argue, planters' phenomenology swept together their lands and their slaves, superimposing non-human and human nature as zones united by their mutual destiny to be dominated and subdued. They saw the land and their slaves as subjects of their authority, despite all of the spectacular evidence – from snow in July to the constant and various resistances practiced by enslaved people – to the contrary. Within this phenomenology, planters could not cede authority to the land lest they concede that enslaved people also escaped total control, an idea that hit on the region's deep anxiety that its slaves would rise up against it.

The belief in human dominance over nature developed over the course of the seventeenth and eighteenth centuries, as the economic and cultural "success" of European colonialism made itself felt.[2] Enlightenment philosophers began to account for this display of European prowess and began to see it as the natural order of things.[3] Their writings, in turn, helped naturalize this attitude. In the introduction to *Systema Naturae* (1735), eminent botanist Carl Linnaeus pronounced

---

[2] See Chaplin, Joyce. *Subject Matter: Technology, Science, and the Body on the Anglo-American Frontier, 1500-1676.* Cambridge: Harvard University Press, 2001 for the compelling argument that it took time and success in the colonies for Anglo-Americans to feel like masters of the earth. During the early colonial period, Chaplin argues, explorers and imperialists felt helpless and endangered, and they looked to native peoples for guidance. In the later period I am discussing, in part due to the continued economic and military success in the colonies, as well as the subjugation of Africans as slaves, white mastery was more an assumption than an open question.

[3] See Foucault, Michel. *The Order of Things: An Archeology of the Human Sciences.* New York: Pantheon Books, 1970.

man the "governor and subjugator of all other beings" (11). According to Hegel, nature's subjugation enabled civilization in the first place. Alienation from nature allowed art and politics to flourish. Peoples constantly at war with the elements were, in Hegel's view, unable to develop culture; to be "close" to nature was to be barbaric. Civilization was, thus, the particular province of Europeans, who were sufficiently free to create culture. Some *philosophes* – like Marquis de Condorcet, who envisioned a time when "the sun [would] shine [on] free men who [would] know no other master than their reason" – even forecasted an age when humanity would have tamed the natural world completely (179). In these thinkers' imaginary, nature was a nuisance that man would, in time, subdue.

Their writings developed what Martin Heidegger might call the period's dominant "ground plan" (*Grundriss*). Heidegger defines a ground plan as the metaphysical premises by which a culture organizes its world (118-123). Heidegger's concept links metaphysics to physics, for these deep metaphysical premises mediate encounters with and interpretations of the physical world, the ground underfoot and the skies above. In historian Michel-Rolph Trouillot's formulation, ground plans dictate what is "thinkable" – or what is perceptible at all (82). In other words, ground plans, as unquestioned assumptions about how the world works, render some phenomena intelligible and interpretable, while excluding or obscuring phenomena that do not conform to underlying assumptions. As the eighteenth century unfolded, human dominance became an integral component of white Westerners' ground plan. Inhabitants of this ground plan were oriented toward the land by it and viewed the land through its lens. Powerful planters, reigning over vast expanses of untapped soil, as well as immense enslaved workforces, seemed to many the realization of Linnaeus's "governor and subjugator of all other beings." As such, they made plans for their ground that accorded to their view of their worldly authority.

Yet, mastery was always an unrealized ideal, a fiction planters labored to sustain. Each plantation was a site of struggle between planter control and the endless challenges posed to it. Environmental historian Lynn A. Nelson defines a farm as an ecosystem that must exclude unwanted elements in order to function (13). Planters attempted to exclude all elements from their lands except their crops and the beings forced to work their fields. This exclusion took the form of a constant battle against invasive elements such as pests, disease, and inclement weather. But antebellum planters were also in the unique position of having to struggle to keep wanted elements – unfree labor – *in*. As invasive elements made their way in, fugitive "elements" made their way out, in lines of flight that planters struggled to intercept. The war to detain fugitives took place in many registers, from legislation such as the Fugitive Slave Acts, to the deputizing of white southerners to create an immense, deregulated police force, to arbitrary and violent "plantation justice," to disciplinary regimes of the Foucauldian sort. Not only did these constant struggles threaten profits, they also threatened to topple the fiction of planter authority, and in so doing struck at the heart of planter self-narrativization and of the most basic assumptions about the natural and social order.

Agricultural discourse in the antebellum South perennially expressed great hope that easy and profitable farming lay just around the westward bend. It is a feature of the discourse to project agricultural paradise – luxuriantly fertile land, cooperative and nurturing weather, and pliant plants – onto the next westward horizon, from Europe to the Old South (Virginia and the Carolinas), finally to the Deep South. In 1773, a Scotsman calling himself "Scotus Americanus" published *Informations Concerning the Province of North Carolina Addressed to the Emigrants of the Highlands*, a pamphlet that sang Carolina's praises. Scotus Americanus reported to his fellow Highland farmers, beleaguered by inordinate rent, debasing landlords, and an unconscionable income gap, that North Carolina was a land of "powerful" vegetation.[4] He indulgently described flavorful fruits, "prodigious quantities of honey," plentiful fish – even "elegant" homes and "sumptuous" entertainment (21-2). Most importantly, Carolina's vegetation ensured that "poverty [was] almost a stranger" in the colony (22). He pleaded with his countrymen to break their "attachment to their place of nativity" and sojourn to the land of plenty (3). Scottish farmers migrated to the Carolina Cape Fear and Piedmont regions in

[4] For a modern history of the social and economic conditions in Scotland described by Scotus Americanus, see Meyer, Duane. "Motives for Migration." *The Highland Scots of North Carolina, 1732-1776.* Chapel Hill: University of North Carolina Press, 1961: 30-53. Meyer's definitive text stresses three major motives for migration: rent hikes, decay of clan structures, and population increases. But he quotes one Bishop Forbes who wrote in 1771, "All, *all*, this is owing to the exorbitant rents for land" (37-8).

larger numbers (between twenty and thirty thousand), Paul Cameron's ancestors among them.[5]

Although upon their arrival the emigrants may have found Carolina's rich alluvial bottomlands and hilly uplands ripe for growing grains and tobacco, their journey to and livelihood within the colony fell far short of the Scotsman's idealized depiction. A century and a half after Scotus Americanus, Southern historian Ulrich Phillips would again praise the South's temperate climate.[6] His infamous opening sentence to *Life and Labor in the Old South* (1929) – "Let us begin by discussing the weather, as that has been the chief agency in making the South distinctive" – displaced responsibility for plantation slavery onto the South's semi-tropical environment and implied that the temperate weather cried out for the institution of plantation slavery (3). But even Phillips acknowledged that the "weather" favored some Southern regions over others. Despite Scotus Americanus's romantic picture, North Carolina was Britain's least profitable colony, in large part, environmental historians argue, because of its terrain. Scottish emigrants would have encountered a rocky, shifting, treacherous coastline and menacing ports. Indeed, Cape Fear, in the colony's southernmost coastal region (where the emigrants would have entered the colony) was so named for its dread-inducing coastline. The area was commonly called the Graveyard of the Atlantic. These coastlines disabled trade by sea, but when colonists looked to overland trade, they found especially poor roadways, as the territory's rivers, running northwest to southeast, thwarted the development of east-west land routes and slowed trade all the more. Unlike Virginia with tobacco, South Carolina with rice, and, later, the Deep South with cotton, North Carolina never claimed a defining cash crop. Conspicuously, the state's largest export was naval stores (tar, turpentine, rosin, and

[5] For a complete history of the Bennehan-Cameron family and lands, see Anderson, Jean Bradley. *Piedmont Plantation: The Bennehan-Cameron Family Lands in North Carolina.* Durham NC: The Historic Preservation Society of Durham, 1985.

[6] See Sutter, Paul S. "Introduction: No More the Backward Region: Southern Environmental History Comes of Age." *Environmental History and the American South: A Reader*. Eds. Paul S. Sutter and Christopher Manganiello. Athens: University of Georgia Press, 2009: 1-24 for a discussion of Phillips's legacy in Southern environmental history. I use Phillips here to suggest that he is in this line of writers who idealized the South's environment, while erasing the labor conditions responsible for any degree of prosperity.

pitch), materials that did not require cultivation but, rather, the destruction of North Carolina's extensive longleaf pine forests.

Historians Kay and Cary describe these environmental features as the colony's "natural handicap." As the nineteenth century progressed, farmers in the Old South compounded these handicaps by overfarming soils that many had come to believe were only moderately fertile in the first place.[7] They practiced what Nelson terms "extensive" farming, or the continuous cultivation of one field without restoring its nutrients, then, when it showed signs of diminished fertility, its abandonment for a new plot (5). Treating the land thus – as disposable – did not grow out of colonists' ignorance about the benefits of restoring old soils (although it took time to develop specific methods for North American soils). As New World traveler Janet Schaw wrote in her *Journal of a Lady of Quality* (1776), émigrés had practiced progressive agricultural methods (what Nelson calls "intensive" farming) in the Old World, where farmers had to be careful not to strip fields that had long been in cultivation. When Schaw visited her brother's Cape Fear plantation, she was dismayed to find that had planters apparently left responsible farming in the motherland. Schaw described extensive farming as a definitively American method, stemming from the sense that land in the New World was an infinite resource and that one need only keep moving west to find fertile and inexpensive land.

By the third decade of the nineteenth century, agriculture in Virginia and the Carolinas was in crisis. The land was commonly believed to be dead. "Extended and superficial culture, with heavy and frequent cropping with tobacco, corn, and wheat, depending wholly upon the native resources of the soil, in time effected an almost total exhaustion of its vegetable and mineral substances, or at least, such a special exhaustion of some of the indispensable ingredients of the surface soil, as to render it incapable of longer producing remunerating crops," wrote farmer James Newman in an article for the *Southern Planter* entitled "Worn-Out Lands in Virginia" (12). The "Valley Farmer," writing in 1858, used the region as a cautionary tale about failing properly to rotate crops: "We need not go from our own coun-

---

[7] Geologically, North Carolina's soils reflect the territory's prior submersion under the ocean. The soils are heavily clay-based and acidic. As I discuss later, amateur soil chemist Edmund Ruffin figured this out and prescribed marl/lime to balance the soil's acidity. In other words, farmers' empirically-derived sense that their soils were not "naturally" overly fertile was correct.

try to find the most lamentable instances of reckless disregard of the correct system of alternation; the evidence is too palpable in all the old States of this Union. Look, for instance, at Maryland, Virginia, the Carolinas and other neighboring States, in which corn, tobacco and cotton, for fifty years or more, have been the leading crops and allowed to follow the same in annual succession upon the same lands until the returns were no longer adequate to sustain their rapidly diminishing population" (492).

Southeastern farmers depleted their soils by using the "three-field shift," planting crops in two fields and using the third as pastureland, in rotation. Under the hooves of hogs and cattle, land was packed, not replenished. In an 1833 issue of the *Farmer's Register*, Virginia farmer Hill Carter called the three-field shift "the most ruinous system that could be invented" (132). Defending his critique in an 1835 issue, he wrote that under the three-field shift, land was "literally grazed to death" (657). Another farmer wrote that this system reduced "originally poor" lands even further, to "sterility" (T.B.A. 612). Farmer James Hamilton Couper agreed, criticizing his neighbors' reckless practices. "When nature is allowed to sow her own seeds and reap her own harvests," he wrote, "the earth, instead of being impoverished by her vegetable productions, seems at each new effort but to augment that fertility." But where farmers coerce the "productions of the soil," soil "exhaustion generally follows, and utter impoverishment [succeeds] teeming fertility" (9).

Despite one anonymous farmer's feeling that "it [was] lamentable and humiliating that it should be necessary to discuss such a topic as the exhaustion of *American lands*; of lands of great natural fertility, which have not been in cultivation, on average, for half a century," farm journals staged heated debates about the state of American soils and the proper care thereof ("Considerations" 605). That the topic of fertilizers dominated these journals indicates a growing awareness that, as a farmer named Bowen wrote in 1849, "constant cropping would speedily exhaust any soil unless we return it to some in some shape or other, those substances of which we deprive it" (329). Those who acknowledged the importance of returning "those substances" that cash crops had depleted debated exactly which substances would work best. Farmers wrote in, recommending their fertilizer of choice: manure, guano, marl or lime, salt, dead animals or their bones and blood, and so on. Or they recommended certain procedures like shade-growing or the planting of a restorative grass (like clover) on a fallow field. These discussions suggest that, in the Old South, the question at hand was whether diminished fertility, felt in diminished prosperity, could be reversed, and how.

But the larger context of this debate was the question of westward migration. Beginning on the 1820s through the next several decades, hundreds of thousands of farmers would abandon their southeastern lands for the new territories, first Alabama and then Mississippi. Agricultural reformers, who believed that poor soils could be amended and restored, pleaded with their brethren not to emigrate, believing that the exodus – and the agricultural attitude that made it necessary – would endanger the South, economically and politically. In an 1852 address to the South Carolina Institute, reformer Edmund Ruffin railed against planters who "deserted the fields they [had] exhausted" only to find "new lands to exhaust in Alabama." In Alabama, they would only do the "work of waste and desolation" over again (481). In this address, Ruffin tempered the imagery he had used thirty years earlier, when he wrote that planters who "spent their lives reducing the fertility of their soils" (through extensive farming), may just as well have cut "twenty throats of each successive generation" ("Morals" 3). A reader of Malthus, Ruffin linked soil exhaustion to depopulation: soil depletion, reduced food sources, and general poverty would cause the next twenty and each twenty thereafter never to be born. And depopulation caused by lower birth rates, as well as depopulation of the Old South by westward migration, led directly, he argued, to the impoverishment of the communities that emigrating farmers left behind. The political and economic stakes of land management could not be higher: to treat the land thus was to endanger the whole region's autonomy, as well as its wealth.

Though Ruffin acknowledged there was a problem with southeastern soils and campaigned vigorously for the application of marl to balance the soil's acidity (an emendation about which, it turns out, he was correct), the true problem was farmers' comportment toward the land. More specifically, the problem was that farmers thought of themselves as *planters* and not as farmers. In this 1836 "Sketch of the Progress of Agriculture in Virginia, and the Causes of Its Decline," Ruffin distinguished between a "planting country" and a "farming country" (26). In the planting country that Virginia once was, planters clear-cut forests, cultivated the land until it was exhausted, and then moved on, westward, clear-cutting again. They did this rightfully, he maintained: extensive farming was the best method in a planting country. But in the settled farming country that Virginia had become, planters could no longer afford to view the land as disposable. As another anonymous farmer wrote in 1848, "Any man with very little skill or knowledge, may fell trees, plant, hoe, cut hay, raise grain, and rear cattle; and so long as his soil lasts, he may do well. At length, however, that becomes exhausted" ("Improvement" 145). Trudging ahead with the practices of the planting country continued the work of exhaustion. Farmers had, in farmer J.B. McClelland's terms, "skinned" the land (253); Ruffin preferred the adjective "land-killing" to describe the misapplication of the planting country mentality to farming country. Both metaphors invoked the violence of the throat-cutting land-tyrants Ruffin had sketched in his early writings. In this way, Virginia and Carolina's farmers were caught between a rock and a hard place, subject to mismanaged and unproductive soils at home, on the one hand, and subject to accusations of political and moral bankruptcy if they migrated, on the other.

In the context of this crisis, representations of an agricultural utopia on the westward horizon again lured distressed farmers to the newest backcountry. In a report that newspapers from New York to Baltimore to Raleigh reprinted for years following its original publication in Alabama's *Halcyon*, one enthusiast described Alabama as "wild and salubrious." The writer took a page from Scotus Americanus's book when he wrote that the "productions of the soil are happily adapted to every species of vegetation," and the "fruit [is] in as great abundance and perfection as in any part of the world." Moreover – and here was the promise of ease as well as profit – vegetation grew "almost spontaneously." The Scotsman had depicted an Edenic Carolina, a picture that, by then, diverged sharply from poor farmers' reality; the *Halcyon* writer promised this unrealized paradise just a little further west. He concluded: "In every point of view that the country can be taken, none on earth presents greater advantages" ("Alabama Territory" 1). An 1835 dispatch from Greene County, Alabama, where Paul Cameron would buy his land, described the area's soils as "remarkable for their health." The writer confirmed that Greene's "inhabitants enjoy a singular exemption from intermittent and bilious fevers, even when the people on the surrounding sandy soils, particularly on the rivers, suffer severely" (Withers 637). In these renderings, the backcountry appears free of the old country's troubles, including uncooperative soils and ravaging disease. Furthermore, these lands are represented as a zone of exception from what farmers knew to be true about the difficulty of cultivation. These renderings erase the distress of the old country with the hopeful promise of the new.

Deep South soils were not only considered exceptionally fertile, they were also considered especially conducive to the production of cotton. One *Farmer's Register* writer calling himself "A Planter" contended that nowhere could cotton be more profitably grown than between thirty-one and thirty-five degrees latitude – Alabama's exact latitudinal boundaries. The Caribbean was too warm (though perfectly suited to sugar production) and Carolina too cold. But Alabama's climate – or what Phillips would later call the "weather" – "perfect[ed] the cotton pods." These pods "yielded [to] the slightest touch of the picker" (418). Economic historians Rhode and Olmstead argue that Southern planters developed new seed varieties, called "Mexican cottons" after their place of origin, that were more resistant to disease and insects, a technology that they claim was especially advantageous to the cotton industry's rise (1123-1171). The *American Farmer* indeed took note that Mexican cottons were "found to ripen in a greater degree at one time, to have a shorter season than any other cotton known among us, and [to be] collected in much greater quantities, by a laborer, in one day." And Mexican cotton, as "A Planter" promised, practically picked itself: "The cotton [hangs] out of the pods of this species of the plants, and even [drops] at times from them." In this romantic picture, supple and marvelous plants would even grant ease to enslaved laborers as they granted huge profits to the land's owners.

It was in this discursive context that Paul Cameron, who will serve as a case study for the remainder of this essay, purchased his Greene County, Alabama, plantation in 1844. Cameron was not deaf to reformers' calls to stay and rehabilitate land in the southeast. Edmund Ruffin was his distant relative by marriage, and Cameron undertook the expensive and laborious (for the enslaved people who executed it) task of marling his Piedmont soils. Indeed, Cameron held onto his nearly thirty thousand acres in the Durham, North Carolina, area and his nearly nine hundred slaves through the Civil War. His profits from the corn, wheat, tobacco, and fruit grown at his Stagville complex were immense; by 1860, Cameron was the wealthiest man in the state. But fifteen years earlier, as a young man under the wing of his distinguished father, Cameron was swayed by what Ruffin called the "madness for emigration" and speculated land in Alabama, where he envisioned ruling over a lucrative cotton plantation ("Sketch" 26).

The dialectic of hope and anxiety was at play from the very beginning of Cameron's transaction. Thomas Ruffin, the well-known North Carolina State Judge and Cameron's father-in-law (to whom I will return at the close of this essay), had reported to Cameron that Colonel William Armistead's "improved" (meaning cultivated) plantation in Greene was "worth the price in weighed gold" – perhaps a reference to the "white gold" whose market value would be determined by its weight. "Had I your means and labour, I'd buy it d___d quick," he counseled (Cameron to Duncan Cameron, 2 December 1845). Cameron bought 1,680 acres from Armistead in October 1844, paying $17.50 per acre for a total of $29,400, and negotiating the price down by offering to pay in cash (Cameron to Duncan Cameron, 27 November 1844).[8] Shortly after making this purchase, Cameron received a letter from his father that warned him against the very transaction he had just made. "I did not receive your letter / the date of which I cannot remember / advising me to rent land until after the purchase had been made," he wrote. This mishap filled him "with much anxiety," but it was "too late to look back." All he could hope was for the plantation to "realize" its "representations" and to "remit" his father (Cameron to Duncan Cameron, 7 November 1844, original emphasis). His father was quite anxious about his untested son's scheme. Cameron's uncle noted in a letter to his brother-in-law: "I trust [Paul] has made for you a desirable investment & I am much relieved, well knowing that it has cost you much anxiety" (Thomas Dudley Bennehan to Duncan Cameron, 9 December 1844).

Affleck's Southern Rural Almanac describes January in Alabama as the time for "cleaning off the land intended for the crop the coming season; fencing, hedging, clearing new land, hauling out manure, &c." (48). So in October, Cameron prepared his slaves for the long march to Greene for their January start to the growing season. He hired an overseer, Mr. Laws, to "take [his] people out," despite feeling disgruntled that Laws charged $140 in fees and that it would cost him "$10 for each horse and 6 or 7 Dollars for each negro" to make the journey

---

[8] This cash offer is astonishing. Most planters paid for their plantations with credit and, thus, accumulated vast debt. That Cameron paid in cash signals how wealthy his family was and how well operations were going back in North Carolina.

(Cameron to Duncan Cameron, 25 October 1844). In early November he helped "reduce things to a good system" and left his slaves "well organized for the March and the encampment" (Cameron to Duncan Cameron, 5 November 1844). At the end of that month, Cameron traveled (by carriage) with thirty-eight enslaved men and boys, thirty-four enslaved women and girls, and twenty-nine enslaved children (on foot) from North Carolina to Alabama, a forced migration that ruptured kin groups that had been established in the relatively insular slave community at Stagville.[9] Cameron reported "get[ting] [his] people to the gate of their new Home" on 7 December 1844 and hiring an overseer whom he initially calls "Mr. Lewelling" – Charles Lewellyn, the overseer at Green from 1845 until late 1849. With great hopes that he would "get it all back," Cameron returned to North Carolina in January, and his enslaved people began preparing the land (Cameron to Duncan Cameron, 7 December 1844).

And so began a steady correspondence between Cameron and his long-distance overseers, first Lewellyn, then John Webster (1849-57), then Wilson Oberry (1857-72), as well as William Lamb, who oversaw the later Tunica, Mississippi, plantation starting around 1859. It should be noted that these letters' "facts" cannot be taken simply at face value. As communications across a power divide between a boss and middle managers, they are discursive constructs, and overseers held an economic interest (the retention of their jobs) in how they represented the plantation. For example, William Lamb repeatedly described all his slaves as "humble" and "submissive." Yet, during Lamb's employ, representatives of C.M. Apperson & Co. wrote to Cameron that they had taken "three of [his] Negroes out of Jail in Memphis, Lem, Meredith, & Evaline" (Apperson to Paul C. Cameron, 11 July 1859). These three had fled Lamb's supposedly subdued plantation. He eventually wrote that he hoped he would "have no more running away" (Lamb to Cameron, 9 July 1859). Ideology shapes remarks by overseers about slaves, representing overseers' desire for slaves' submission. Their comments about agricultural output also cannot be taken as pure statements of fact, as overseers had an abiding interest in how they represented their crops. That said, agricultural output was quantifiable in a way that slave resistance (other than out-

[9] For genealogies of Stagville's enslaved community, see McDaniel, George. *Stagville: Kin and Community*. Durham NC: Published by author, 1977.

right fugitivity) was not, at least not to overseers. Agricultural output was measured in the acreage plowed, the number of bales produced, and the prices the crops fetched in Mobile. Overseers' narratives can provide some account of the "actual" successes and failures at Greene, while also providing a discursive account of how they perceived these plantation and the distress these events caused them.

What transpired at Greene was a series of agricultural catastrophes. The kind of unruliness overseers witnessed at Greene was distinctly different from the kind of unruliness they had left behind in the "dead" southeast, discussed above. There, overfarming had exhausted the land, and now the land seemed to resist their agricultural efforts. But Greene's unruliness was spectacular. During the first growing season (1845), Lewellyn's hope for a "fair crop" gradually gave way to a sense that the cotton crop was "indifferent" (Lewellyn to Cameron, 1 June 1845 and 3 July 1845).[10] In the late summer, Lewellyn realized that the cotton crop was even worse than indifferent. He realized he would bring to market just a fraction of the expected amount because disease, "the rust," was attacking his crop: "the rust [had] nearly ruined [it]" (Lewellyn Cameron, 1 August 1845). The rust Lewellyn described was probably cotton leaf rust, caused by the fungus *Puccinia cacabata*. Cotton leaf rust causes bright yellow spots on the plant's leaves, bracts, and bolls and renders the plants unusable. In addition to the rust, Lewellyn also notes that lice and "the worm" – likely either the boll worm or the cotton worm – infected the crop. (Unbeknownst to Cameron and Lewellyn, after decreasing in the 1830s, these worms increased significantly across the Deep South during the 1840s.) When his "indifferent" slaves began harvesting in the middle of August, Lewellyn lamented that he was "not getting nothing like what [he] expected." The cotton, he wrote, had "not grone any since the 13th of July [and] the rust [was] all over the plantation" (Lewellyn to Cameron, 16 August 1845).

Then came the rains. Hard rains late in the cotton growing season washed unharvested cotton away. Rains also "stained" (made moldy)

[10] Lewellyn uses this same word ("indifferent") repeatedly to describe the efforts of the enslaved laborforce at Greene as well as the crops, a point to which I will return. Here, I note that Lewellyn's repeated use of the words links the land with the slaves as related impediments.

cotton bolls that could, in turn, ruin neighboring plants. In October, Lewellyn wrote, "it commenced raining here yesterday morning and is now raining, and the ground is completely covered with cotton, washed out by the rain. I have picked no cotton within the last four or five days" (Lewellyn to Cameron, 8 October 1845). Lewellyn's distress is underscored by his syntax, his clauses propelled by a series of "ands": it is raining, and it won't stop raining, and the crop has been washed out, and there is nothing to be done. Indeed, brutal rains constitute a refrain throughout overseers' letters. "We had a rain here on Monday evening that like[d] to have washed away every thing[.] I shall have to plough up the land on the creek and plant over in cotton, [as] my stand is very mutch injured," reported Lewellyn in April 1846. That summer, the cotton crop was "completely ruined by the worm and rain" (Lewellyn to Cameron, 30 August 1846). In the 1847 growing season, insects (caterpillars, grasshoppers, lice, and worms) and rain invaded. Lewellyn wrote in July: "your crop of cotton is very mutch injured by the rain on Thursday[;] it is wash[ed] up and covered up in mud. I never saw such a rain fawl as fell on that day" (Lewellyn to Cameron, 10 July 1847). That year, Cameron explained to a Mobile merchant that the plantation would make "little or no crop of cotton" (2 October 1847).

Greene's dazzling unruliness stimulated an exaggerated discourse of affect in Cameron's writings. Hopefulness turned to anxiety, which, after the first growing season, turned to "mortification." Cameron describes feeling "sorrow" and "anger" for having made himself an "object of derision" in the eyes of his father, and he blames his father-in-law, Thomas Ruffin, for falsely representing the land. He regards himself "not only as a dupe but as a victim" (Cameron to Duncan Cameron, 18 November 1845, original emphasis). A similar distress pervades overseers' letters. "It has bin raning nearly all the weak and looks to me it is going to rane all tha year," wrote Wilson Oberry to Cameron in 1859. The unintended homonym produces a double meaning: "it has been raining all week" and "it has been raining, and all are weak." His exaggeration ("all tha year") captures his feeling of being drowned by the elements. Lewellyn claimed in 1847 that he was "never so completely worn out with sickness in [his] life" (Lewellyn to Cameron, 28 September 1847). The repeated use of superlatives suggests that overseers were stretching to find language adequate to their emotional states as they attempted to explain to Cameron what it was like to live under Greene's unruliness.

Overseers perceived events at Greene as surreal. Recall Webster's description of snow in July as "strange whether." The construction of Greene's environment as surreal is perhaps most clear in an 1859 exchange between Cameron and Oberry. Cameron wrote to inform Oberry that some Mobile merchants had sent along positive reports of Greene's cotton. In turn, Oberry requested Cameron "be so good as to send [him] them letters them gentlemen sent [him] last year[.]" He continued: "i will knot make any fus about it[;] i mearly want to know hu tha are" (Oberry to Cameron, 12 March 1859). Oberry's request for the material documents suggests that he desires more than simply to know "hu tha are." In the surreal world of Greene, during "one ove the hardist years [he] ever saw to get along with bisnis," Oberry requests the actual letters so that he might know with certainty that they are real. Paper becomes the compensatory object that, unlike the plantation, will not resist being held.

Given that Greene apparently caused Cameron such emotional strife, it might seem surprising that, upon Greene's failure, Cameron simply removed further west, to Tunica, Mississippi. But, in light of my discussion of the agricultural promises of the west, we should not be surprised. Although hope for the agrarian utopia was responsible for farmers' sense of despair at the reality "on the ground," this despair did not stop planters from ritualistically looking further west, there to project the seeds of a new hope and a new disappointment. The terms of this dialectic were underwritten by planters' imperative to self-identify as powerful proprietors and rightful masters. Drawing on Bourdieu, Trouillot defines the "unthinkable" as "that which one cannot conceive within the range of possible alternatives, which perverts all answers because it defies the terms under which the questions were phrased" (82). Cameron, like his fellow westward-marching planters, disabled himself from thinking the reality that soils and plants did not inevitably obey the authority of the planter.

Cameron performs this disavowal rhetorically when, in 1846, he describes Greene as the "**theatre of [his] sacrifice**" (Cameron to Duncan Cameron, 18 November 1845, original emphasis). To call Greene a "theatre" in the first sense of the word, a scene of staged drama, is to construct it as a space where illusions play. In the metaphor, Greene becomes a zone where fictions unfold and where onlookers to the event are merely spectators who will, in time, leave the illusions behind and return to the real. This construction turns Greene's figuratively unbelievable physics into a kind of literally unbelievable physics according to which events do not have to be believed because they are not, in fact, real. Constructed initially as an exceptional, utopian space ("worth the price in weighed gold," "remarkable for its health"), Cameron, upon his disappointment, first understood these descriptions as false representations. But imagining Greene as a theatre allows Cameron to position "reality," or the actual events at Greene, in the zone of representation, as well. Nothing at Greene is real. The metaphor's sense of enclosure allows Cameron to draw a ring around Greene and consider it exceptional in a second sense: an exception from the real, a bizarre, contingent zone cut off and erased from experience.

"Theatre" also signifies as a demarcated zone of conflict in a larger war. America's western frontier becomes its western front in a war against natural elements (and native peoples, as well). In the sense that Alabama was a theatre of war, the plantation becomes the scene of a battle lost in a war that can still be won. The metaphor constructs Greene's natural elements as actors in insignificant skirmishes whose victory does not say anything about the actual power of human force. Naming the conflict a "sacrifice" extends the metaphor by suggesting a volitional giving over, a strategic defeat for the sake of the larger cause (giving up Greene for dominance elsewhere). In sum, the metaphor reveals Cameron constructing Greene as neither real nor decisive and absorbing its resistance into his worldview as nothing more than an aberration.

Hope restored, Cameron carried on as before. As early as 1848, he began speculating land in the *new* new backcountry of Mississippi. He argued to his father that his mistake at Greene was to have bought improved land when he should have bought uncleared, forested, nutrient-rich land. He told his father that he ached for Mississippi's "virgin soil with its fruits." His use of "virgin soil" not only reiterates agricultural discourse's romanticization of the unknown frontier, but it also recalls descriptions made at the "discovery" of the New World, casting Mississippi as the bountiful, virgin American continent of splendorous, yet-unknown riches. Moreover, when Cameron wrote that land in Mississippi was "so clean that a 2 horse plough would go along on it as old land – the roots grow so deep in the soft rich soil," he imagined the soils as both virgin *and* old, containing all their original, untouched nutrients but being workable like the old land trained to yield under the plow. Cameron was so carried away by the thought that he spelled out a detailed plan to his father. He imagined relocating all of the family's slaves and resources to Mississippi except those needed to keep some grains and clover growing back in Carolina, where Stagville would become what Frederick Law Olmsted termed a "show plantation." In his eagerness, Cameron declared that "should [his father] approved of this thought [he could] have the emigrants (slaves) shad and clothed & under way in 5 days from the time [he went] about it" (Cameron to Duncan Cameron, 2 November 1848, original emphasis). But his father did not approve.

He would have to wait until 1856 to move ahead with his plan. That year, he used his inheritance to buy the Mississippi land he had eyed eight years prior. Buying unimproved land in Tunica, Mississippi, from Sam Tate, he again prepared "his people" to march. Greene's overseer at that time, John Webster, marched slaves north-westward in December of that year. Upon his return to Greene, Webster wrote Cameron that he had bought "as rich land as [he] ever saw" but warned him that "it will [be] a great deal of labor, time, and money" to clear it (Webster to Cameron, 24 December 1856). To oversee the initial clearing, Cameron hired William Lamb, under whom Tunica was deforested and ditched, plowed and planted, and the Mississippi land began to give Cameron what he had been seeking for a decade: a fruitful cotton plantation. Cameron celebrated his long-awaited reward by writing in 1859 that he was "more and more pleased with the property" and that he "shall in five years have a large increase from it if nothing happens" (Cameron to Thomas Ruffin, 3 January 1859).

In his 1861 documentary account of the South, *The Cotton Kingdom*, Olmsted showed that the cotton boom benefited only the biggest, most established planters, who bought up large tracts of the most fertile soil, boxing out smaller planters from good land and driving up the cost of slaves, whose labor was in high demand but in short supply. Big planters composed just a fraction of white southerners farming the Deep South, but the economy they produced made it nearly impossible for ordinary farmers to profit. "Except in certain limited districts, mere streaks by the side of rivers and in a few isolated spots of especially favored soil," Olmsted wrote, "citizens of the cotton States, as a whole, are poor" (13). Drawing on De Bow's census and anticipating modern economists' assessment that plantation slavery was in fact economically unsound, Olmsted did the math to estimate that ordinary cotton farmers, even in an "extraordinary" year, would profit around one hundred and twenty-five dollars off cotton, all of which would have the be reinvested in the day-to-day operations of the farm (18). If they did not diversify their crops with corn or grains, these smaller farmers would live out their lives in the red. Both anecdotally and quantitatively, Olmsted concluded that "a majority of those who sell the cotton crop of the United States must be miserably poor – poorer than the majority of our day – labourers at the North" (19).

But Southerners continued to migrate with the belief that nothing, either natural or political, would "happen." That nothing should happen was a fiction, iteratively uttered, as the necessary corollary to the fiction of planter mastery. It had to do with how planters saw themselves and how they saw the world around them. More specifically, it had to do with planters' efforts to construct and make true a kind of relation between themselves and, in Linnaeus's terms, "all other beings." Within this view, these others had to become non-happenings, or beings with absolutely no potential for happening. All other beings had to be stripped of their wills, and the language of utopianism this essay has traced was one rhetorical way planters enacted this ideal relation: the utopianism of tamed nature.

But the "other beings" truly at the center of this ideological project were enslaved peoples. Nowhere is this point clearer than in the *North Carolina v. Mann* (1829) opinion, written by Thomas Ruffin, Cameron's father-in-law. The opinion gave slaveowners "uncontrolled authority over the bod[ies]" of their slaves. Ruffin's reasoning was as follows: slavery is only safe and profitable if the master has "absolute" power, and for the master to have absolute power, the slave's "submission" must be "perfect." This perfect submission can "only be expected from one who has no will of his own." Masters must be handed unbridled authority so as to effect the complete shattering of slaves' will. Yet, this opinion contains a paradox that reveals its status as (for masters) a utopian fiction. The opinion imagines the perfect submission of the slave, the complete stripping of his will, while also authorizing masters' absolute power precisely *because* of slaves' insistent willfulness, or their ever-imperfect submission. Indeed, were slaves' "perfect" submission possible, the *North Carolina v. Mann* decision would be wholly unnecessary. In this paradox, Ruffin's opinion betrays itself as a performative fiction, attempting to write into existence an ideal it already doubts.

It is in this context that we can interpret the dialectic of hope and distress that stretched planters across the South. This essay has traced the renewal of hope at each disappointment as a disavowal that put under erasure what each unruly environment threatened to reveal, namely that humans cannot perfectly master natural environments. Instead of letting this lesson emerge, planters like Cameron simply projected westward, chasing the ideal inherent to the dominant ontology that centered them as rightful masters of the earth.

Because a relation of "absolute authority" lay at the heart of planter metaphysics, land and slaves were grouped together under the heading of beings meant ("naturally") for domination and work. Perfect submission was required of both non-human and non-white human zones. Displays of willfulness in either zone could not be countenanced by planters because such a recognition would disrupt the fundamental position of planters in relation to all other beings. In other words, to countenance imperfect submission, or uncontrollability, in other beings would call into question the most basic assumptions about planters' place in the world and, in turn, the social structures built thereupon. The disavowal caught up in the westward utopianism served a similar function to the image in *North Carolina v. Mann* of the slave without a will. The disavowal props up the central fiction of masters' authority, staving off the unthinkable: that both the natural world and enslaved people had the capacity to defy that authority.

Tracing the affective language and moods in agricultural discourse, as well as the "actual" failures of the plantation, however, shows, as Ruffin did in *Mann*, that farmers cycled through their own doubts concerning the completeness of their authority. But these doubts remained pinned in the affective register, for, again, to think or speak some other relation would seem to jeopardize their dominance over their slaves. Although it was felt, distress, or the possible meanings and implications of distress, could not be thought. Hence, hope prevailed, and planters trudged westward, gripping tightly to the fiction of nature's perfect non-happening, their enslaved people forcibly in tow.

Works Cited

A Planter. "On the Sugar and Cotton Regions of the United States." *Farmer's Register*. October 1838: 418. Print.

Affleck, Thomas. *Affleck's Southern Rural Almanac, and Plantation and Garden Calendar*. Washington MS: Thomas Affleck, 1852. Print.

"Alabama Territory, From the St. Stephen's *Halcyon*." *New York Spectator*. 16 September 1817: 1. Print.

Americanus, Scotus. *Informations Concerning the Province of North Carolina Addressed to Emigrants of the Highlands*. Glasgow: Printed for James Knox, 1773. Print.

Anderson, Jean Bradley. *Piedmont Plantation: The Bennehan-Cameron Family Lands in North Carolina*. Durham NC: The Historic Preservation Society of Durham, 1985. Print.

Bowen. "Keeping Up Fertility." *Southern Planter*. November 1849: 329. Print.

Carter, Hill. "The Four-Shift System: The Best Rotation for James River Lands." *Farmer's Register*. August 1833: 132. Print.

—. "Remarks on the Comparative Advantages of the Three-Shift and Four-Shift Rotations." *Farmer's Register*. April 1835: 657. Print.

Chaplin, Joyce. *Subject Matter: Technology, Science, and the Body on the Anglo-American Frontier, 1500-1676*. Print.

Condorcet, Jean-Antoine-Nicolas de Caritat. *Sketch for the Historical Picture of the Progress of the Human Mind*. New York: Noonday Press, 1955. Print.

"Considerations on the Causes and Effects of the Diminution of American Crops." *Southern Planter*. October 1859: 605. Print.

"Cotton." *American Farmer*. 7 July 1820. Print.

Couper, James Hamilton. "Essay on Rotation of Crops." *Farmer's Register*. June 1833: 9. Print.

Foucault, Michel. *The Order of Things: An Archeology of the Human Sciences*. New York: Pantheon Books, 1970. Print.

Hegel, George F.W. *Lectures on the Philosophy of History*. Oxford: Claredon Press, 2006. Print.

Heidegger, Martin. "Age of the World Picture." *The Question Concerning Technology, and Other Essays*. New York: Harper &Row, 1977. Print.

Kay, Marvin L. Michael and Lorin Lee Cary. *Slavery in North Carolina, 1748-1775*. Chapel Hill: University of North Carolina Press, 1995. Print.

Linnaeus, Carl. "Introduction to *Systema Naturae*." *Race and the Enlightenment: A Reader*. Ed. Emmanuel Eze. Cambridge: Blackwell Press, 1997. Print.

McClelland, J.B. "Tobacco, not Necessarily an Exhausting Crop, and no Demoralizer." *Southern Planter*. April 1859: 253. Print.

McDaniel, George. *Stagville: Kin and Community*. Durham NC: Published by author, 1977. Print.

Meyer, Duane. *The Highland Scots of North Carolina, 1732-1776.*Chapel Hill: University of North Carolina Press, 1961. Print.

Nelson, Lynn A. *Pharsalia: An Environmental Biography of a Southern Plantation, 1780-1880*. Athens: University of Georgia Press, 2007. Print.

Newman, James. "Worn-Out Lands in Virginia." *Southern Planter*. March 1852: 12-3. Print.

Rhode, Paul W. and Alan L Olmstead. "Biological Innovation and Productivity Growth in the Antebellum Cotton Economy." *Journal of Economic History* 68:4 (2008): 1123-1171. Print.

Olmsted, Frederick Law. *The Cotton Kingdom: Traveler's Observation on Cotton and Slavery in the American Slave States.*New York: Mason Bros., 1861. Print.

"On the Improvement and Management of Soils." *Southern Planter*. May 1848: 145. Print.

Phillips, Ulrich. *Land and Labor in the Old South*. Chapel Hill: University of North Carolina Press, 2007. Print.

Ruffin, Edmund. "An Address: On the Opposite Results of Exhausting and Fertilizing Systems of Agriculture, Read Before the South Carolina Institute at its Fourth Annual Fair, November 18, 1852." *Southern Planter*. August 1860: 481. Print.

—. "The Morals of Agriculture." *Nature's Management: Writings on Landscape and Reform,1822-1859*. Ed. Jack Temple Kirby. Athens: University of Georgia Press, 2000. Print.

—. "Sketch of the Progress of Agriculture in Virginia, and the Causes of its Decline." *Nature's Management: Writings on Landscape and Reform, 1822-1859*. Ed. Jack Temple Kirby. Athens: University of Georgia Press, 2000. Print.

Schaw, Janet. *Journal of a Lady of Quality; Being the Narrative of a Journey from Scotland to the West Indies, North Carolina, and Portugal, in the Years 1774-1776*. Lincoln: University of Nebraska, 2005. Print.

T.B.A. "On the Condition of Farmers on Poor Lands." *Farmer's Register*. March 1835: 612.

Trouillot, Michel-Rolph. *Silencing the Past: Power and the Production of History*. Boston: Beacon Press, 1995. Print.

Valley Farmer. "Alternation of Crops." *Southern Planter*. August 1848: 492. Print.

Withers, Robert W. "Some Account of the Calcareous Region of Alabama." *Farmer's Register*. March 1835: 637. Print.

Temptation of Fruit: The Symbolism of Fruit in Christina Rossetti's "Goblin Market" and in the Works of D. G. Rossetti and J. E. Millais

Akemi Yoshida

Abstract: Though critical writing often refers to biblical allusions in "Goblin Market" and to its association with Milton's *Paradise Lost*, strangely little has been written concerning its relation to contemporary fruit images, which actually seem to have permeated mid- and late Victorian culture. This paper explores the symbolic meanings of fruit as it appears in the works of the Rossettis and Millais, and attempts to prove that while sharing the same symbolic codes as her male counterparts, Christina Rossetti tries to nullify the ominous fatality symbolized by fruit as anachronistic, and to liberate her female characters from the mythic roles into which fruit seems to fix women in D. G. Rossetti and Millais's work.

Though attempts have been made to read Christina Rossetti's "Goblin Market" (1862) in non-sexual terms, most critics seem to agree that the "fruit forbidden" which appears in the poem should be understood as a metaphor for erotic attraction and sexual delights. Like Dorothy Mermin, who rejects readings of this poem as the unconscious revelation of a duality within the poet, this paper assumes that Rossetti consciously employed the metaphor of "fruit forbidden" to present what would have been otherwise unsayable. Members of the Pre-Raphaelite Brotherhood, including Dante Gabriel Rossetti and John Everett Millais, often employed images of fruit in their prose or poetic works and in their paintings which carried sexual connotations, and it would be natural to suppose that Christina Rossetti was familiar with such symbolic usage of fruit. But are there any differences in the usage of the fruit imagery in "Goblin Market" from that in the works of Dante Gabriel Rossetti or Millais? Though critical writing on "Goblin Market" has often referred to its biblical allusions and its association with Milton's *Paradise Lost*, strangely little has been said in terms of its relation to contemporary fruit images despite the fact that such images actually seem to have permeated mid- and late Victorian culture. This paper explores the symbolic meanings of fruit as it appears in the works of the Rossettis and Millais, and attempts to show that while sharing the same symbolic codes as her male counterparts, Christina Rossetti tries to nullify the ominous fatality symbolized by fruit as anachronistic, and to liberate her female characters from those mythic roles into which fruit seems to fix women in D. G. Rossetti and Millais's works.

I

"Goblin Market" narrates a story of two sisters, Laura and Lizzie. "Morning and evening," they hear "the goblins cry: 'Come buy our orchard fruits, / Come buy, come buy.'" Though Lizzie admonishes her not to hear or look at them, Laura takes interest in the fruit offered by the goblins, "wonder[s]" at the multifarious appearance of the goblins, and at last, goes to buy the fruit from them. After tasting the juicy fruit once, however, Laura, though desiring to buy their fruit again, no longer can hear the goblins. The goblins' fruit, which is "Sweeter than honey from the rock" (l.129) and whose juice is "Clearer than water" (l.131), seems to generate in Laura a thirst which never can be satisfied and which erodes her life from inside. "Her hair grew thin and gray; She dwindled, as the fair full moon doth turn / To swift decay and burn / Her fire away" (ll.277-280). The motif of poisonous fruit, beautiful to the eye but fatal to the one who tastes it, might remind the reader, for example, of the apple in Brother Grimm's "Snow White." Indeed, it has been pointed out by critics such as Terrence Holt that the narrative style of "Goblin Market" is that of a fairy tale (195). And as with other traditional fairy tales, some critics have tried to draw a moral lesson from the sisters' story. Gilbert and Gubar, for example, interpret Laura's suffering as punishment for women's attempt at self-expression. Is it, however, appropriate to take seriously Laura's story as Rossetti's warning message addressed to her female contemporaries? Admitting that the rich symbolism of Rossetti's poetry allows multiple interpretations and should not be reduced to a single, monolithic reading, I would prefer to regard the use of fairy tale narrative and the symbolism of fruit as a parody, rather than as a serious application and adaptation of these tropes to the mid-Victorian cultural context.

The effect that eating the goblin fruit has upon the female character is clearly a curse, but it should be noted that no specific moral reproach is given regarding the act of buying and consuming the goblin fruit. The goblin fruit is to be avoided, because it seems to promise the ruin of the consumer, but the assumption that the goblin fruit inevitably brings about its "cankerous" effects is based on the deduction from the precursors' destiny, rather than on any moral ground:

> Dear, you should not stay so late,
> Twilight is not good for maidens;
> Should not loiter in the glen
> In the haunts of goblin men.
> Do you not remember Jeanie,
> How she met them in the moonlight,
> Took their gifts both choice and many
> Ate their fruits and wore their flowers
> Plucked from bowers
> Where summer ripens at all hours?
> But ever in the noonlight
> <u>She pined and pined away;</u>
> <u>Sought them by night and day,</u>
> <u>Found them no more but dwindled and grew grey</u>
> Then fell with the first snow,
> While to this day no grass will grow
> Where she lies low:
> I planted daisies there a year ago
> That never blow.
> You should not loiter so." ("Goblin Market", ll. 143-156)

When Laura comes home late after sucking the goblin fruit, Lizzie "upbraid[s]" her, citing the example of "Jeanie." Jeanie is another girl who, after taking the goblin fruit, "pined and pined away, /sought them by night and day, /Found them no more but dwindled and grew grey" (ll.154-156) and died. Laura, not seriously heeding her sister's advice, claims the benefit she has derived from the goblin fruit; that is, she has "done with sorrow" after eating the fruit. Nevertheless, she also has already fallen into the same rut as Jeanie by this time and is to go so far as to knock at "Death's door" (l.13). Lizzie, though wanting to "buy fruit to comfort [her sister]," is discouraged to do so by the fear of paying "too dear," because:

> She thought of Jeanie in her grave,
> Who should have been a bride;
> But who for joys brides hope to have
> Fell sick and died
> In her gay prime.

When Lizzie finally determines to take action to rescue her sister, however, she is described as a being who, "for the first time in her life/ Began to listen and look" (ll.327-8). The myth of Jeanie has such a disempowering effect that it has hitherto succeeded in depriving Lizzie of her ability to "listen" or "look." Phrases such as "joys brides hope to have" of course lead the reader to associate the goblin fruit with sexuality, but the fear of the curse of the fruit suppresses in females not only their sexual interests, but also their capacity to grasp the reality through their perceptions.

The phrase "fruit forbidden," which is used to refer to the goblin fruit, clearly connects the world of "Goblin Market" to that of *Genesis*, incorporating Jeanie and Laura into the genealogy of Eve. Naoko Yoshida points out the importance of paying attention not only to the relation between "Goblin Market" and *Genesis,* but also to that between "Goblin Market" and Milton's *Paradise Lost*, in which emphasis is put upon the acquisition of knowledge as the result of the act of Eve's eating the fruit. True, there are similarities between "Goblin Market" and *Paradise Lost* in the presentation of the rich imagery of fruit which suggest that Rossetti has drawn upon Milton's text. However, when exploring the significance of fruit imagery in "Goblin Market," it is necessary not only to look at the direct relation between Rossetti's text and classical texts such as *Genesis* and *Paradise Lost*, but also to note that the symbolic use of fruit was actually prevalent in the works of Pre-Raphaelite artists such as Dante Gabriel Rossetti and John Everett Millais.

II

It has often been pointed out that the Pre-Raphaelites, though rejecting classical idealism in depiction, preferred to use medieval or mythical settings. In her *Women/ Image/ Text: Readings in Pre-Raphaelite Art and Literature*, Lynne Pearce has pointed out the "necessity of myth," which drove many Pre-Raphaelite artists to repeatedly depict the "Lady of Shalott," whose image is overlapped with that of middle-class women of the Victorian period, for whom it was forbidden to go out of the household to get in touch with the reality outside. Pearce also analyzes the danger of the symbolism in such works as Dante Gabriel Rossetti's painting *The Girlhood of Virgin Mary* and his sonnet attached to it. D. G. Rossetti's sonnet instructs the viewer how to read each item in his painting as a symbol which underlines the virtue and the "perfect purity" in Mary. According to Pearce:

> [I]nstructions necessitate rules, and until such a time as she can position herself to break free of them, the viewer who has swallowed Rossetti's painting thus whole, will necessarily be their victim [...] Symbols are 'read' as explanations [...] By such means is the web of cultural ideology extended; by learning discourses we learn to obey them. (34)

In the works of Dante Gabriel Rossetti and John Everett Millais, fruit is often used as a convenient adjunct which suggests historical and mythical palimpsests. Fruit has traditionally been associated with sexuality (at least since *Genesis)*, and I would like to focus on two of its symbolic functions: how the act of offering and taking the fruit represents sexual negotiation, and how the fruit itself is often identified with the female body.

The motif of a woman's taking fruit from a man has frequently been employed with specific symbolic purposes. In Greek mythology, Proserpine's tasting of the pomegranate binds her irretrievably to the world of Pluto. Once tasted, the fruit cannot be restored: the image of fruit functions here as a metaphor to present the irrevocability of a step taken in sexual negotiation. Millais's *The Woodman's Daughter* (1851) is a painting based on a poem by Coventry Patmore that narrates a girl's tragic fall as the result of her love to a man of different social status. This painting depicts the moment of the transference of strawberries from the "rich Squire's son" to Maud the woodman's daughter as a decisive one. It corresponds to Patmore's lines:

> Her secret's this: In the sweet age
> When heaven's our side the lark,
> She follow'd her old father, where
> He work'd from dawn to dark,

For months, to thin the crowded groves
Of the ancient Manor-Park.

She fancied and he felt she help'd;
And, whilst he hack'd and saw'd,
The rich Squire's son, a young boy then,
Whole mornings, as if awed,
Stood silent by, and gazed in turn
At Gerald and Maud.

And sometimes, in a sullen tone,
He offer'd fruits, and she
Received them always with an air
So unreserv'd and free,
That shame-faced distance soon became
Familiarity.

Therefore in time, when Gerald shook
The woods, no longer coy,
The young-heir and the cottage-girl
Would steal out to enjoy the sound of one another's talk,
A simple girl and boy.

In those lines from the former part of the poem, though they apparently depict an innocent love between young children, there can be already detected an ominous tone heralding the tragic result of the relationship, the girl's having a baby and secretly disposing of it. While the couple are described as "a simple girl and boy," certain connotations of sin are also suggested in such phrases as "Her secret", "shame-faced," and "steal out to enjoy." Robert M. Polhemus points out the parallel between this scene and *Genesis*:

> He [Millais] is re-figuring *Genesis* and Eden and rendering the generation of the Fall of childhood. The sire-gardener-woodman labors away, impotent to protect the daughter … The cold, beautiful, upper-class boy can be read as an original sinner, but possibly redeemable. (298)

Curiously, though both parties have at least equally committed themselves to the mutual relationship, the affair is referred to as only "Her secret." It seems that the narrator of the poem regards her as the guilty one. The boy seems more conscious of what he is doing, watching out for an occasion for approaching the girl, standing "silent by, and gaz[ing] in turn/ so unreserv'd and free," but the consequence falls only upon her.

Millais has specifically chosen for his work the scene where Gerald, the "squire's son," is offering strawberries to Maud, who is about to take them. Gerald and Maud are depicted as having a curiously fixed, two-dimensional impression, lacking depth and totally separated from the background, where the woodman is. The moment of the act of Maud's taking the fruit implies her fall and her destiny hereafter, just as the mischief of Eve, the first woman, has so often been recorded as the archetype of the sin of all women coming after. In Dante Gabriel Rossetti's *Bocca Baciata*, there is an apple placed in front of the woman, the apple of the same red color as her "kissed lips." The title, which means "kissed lips," implies that once kissed, the lips cannot remain in the former state – her eyes are still dreaming in the resonance of intoxication she has just experienced, and the symbolic apple tells that she has been sexually imprinted.

In a poem and a story entitled "The Orchard Pit," D. G. Rossetti depicts the skeletons of men whose taking apples from a *femme fatale* has irresistibly led them to their ruin. Similarly, in his poem and painting *Proserpine* (1873-77), D. G. Rossetti overlaps three female figures belonging to different cultural and historical contexts to merge them into one: Proserpine, Eve, and Jane Morris. According to the episode of "The Rape of Proserpine" in Ovid's *Metamorphoses*, the girl, who has had the "ambition" to "remain a virgin" is taken away by Pluto, and "innocently picked a pomegranate from a drooping branch, and had placed in her mouth seven seeds taken from its pale husk" in his domain, not knowing she "may return to heaven, but on one definite condition, that no food has passed her lips in that other world." In D. G. Rossetti's painting, however, the dominant female figure meditating upon the fatal pomegranate seems well aware of the meaning of her deed, having determined reluctantly to take the responsibility of committing herself to an unwelcome marital relationship.

The image of fruit thus seems to crystallize and mold women into mythical types, deprive them of independent personality, and destine them for irresistible ruin. Christina Rossetti might have detected danger in reproducing such images which would threaten to reduce the women living in the real world to types and make them live under the shadow of their archetypical predecessors. In "Goblin Market," Christina Rossetti first presents the fruits offered by the "goblin men" as fatal, like the apple in D. G. Rossetti's "The Orchard Pit": "Jeanie," who has succumbed to the temptation of their fruit, dies, and Laura, one of the protagonists of the poem, also suffers from a life-threatening bout of consumption. The familiarity of characters' names underlines that they are our contemporaries and actually live among us, giving the impression that the story still holds true. Near the end of the poem, however, the narrator distances the reader from taking the fruit metaphor too seriously by attributing the goblin men and their fruits to a time past: "those pleasant days long gone of not-returning time." The omen of fatal fruits is safely placed in the historical past and will not fall upon the coming generations, including the children of Laura and Lizzie, who are now listening to the tale of the goblin men as if to a fairy tale. The difference in tense between the parenthesized passages and the rest of the poem reinforces this impression. The line, "Men sell not such in any town," which appears twice in the poem and which is the only sentence written in the present tense amid the past tense narrative, seems to emphasize that this fruit myth belongs to a different sphere of the supernatural and mysterious past, and to reassure the reader that this has nothing to do with the real "men" or humans living in the present.

III

Another stereotype women tend to be molded into when juxtaposed to the image of fruit is the stereotype of femininity itself. In Millais's painting *Cherry Ripe* (1879), which achieved extreme popularity among contemporary viewers, the charming girl and the cherries, metonymically identifiable, offer themselves to the viewer. The employment of the singular form – "cherry," rather than "cherries" – in the title of the painting strengthens the symbolic interpretation of the title as directly referring to the girl. Though she is tiny and as sweet as a cherry, she is already "ripe" – her straight gaze unflinchingly tells the viewer of their mutual complicity, her knowledge of the way of the world. *Cherry Ripe* was painted after the publication of "Goblin Market," but similar examples of identification of fruit and a female figure as a sexually consumable object can be also found in Millais's early works such as *The Bridesmaid* (1851).

While taking the fruit brings about the fall of the women, it seems that women are at the same time expected and demanded to take the fruit, to submit to the sexual code. The orange is a symbol of the marital relationship, and in *The Bridesmaid*, the dominant female figure is, according to Tim Barringer, "contemplating with fear and fascination future sexual consummation" with the intact orange in front of her. This is regarded as the normal mental state of a girl yet unmarried, for "Marriage was the most important event in the life of middle-class women in Victorian Britain, a transition from virginity to motherhood, from girl to woman [...] Victorian women not attaining the state of matrimony were considered either unfortunate or eccentric." Her orange hair falls down over her like a cascade, covering her body in the same way that the hair of Mary Magdalen is conventionally painted, and identifies the girl with the orange served on the plate, ready to be tasted.

In "Goblin Market," Lizzie's body becomes identified with fruit as an edible object, but not as a sexual object to be consumed by male desire. It has been pointed out by such critics as Mermin that Laura's act of giving herself up to Lizzie, which works as an antidote to the poison of the fruits of the goblin men, overlaps with the image of Christ – the deliverer who would expiate the sins of all human beings. Laura seems to succeed in freeing herself and other women coming hereafter from negative stereotypes as doomed beings associated with sin.

In a paper titled "The Meretricious and the Meritorious in 'Goblin Market': A Conjecture and Analysis," D. M. R. Bently presents the hypothesis that "Goblin Market" was written to be read to an audience of "fallen women" and to help them reform themselves. Whether this "conjecture" is true or not, it might be said that Christina Rossetti, by differentiating her usage of the fruit image from that of D. G. Rossetti or Millais, and has attempted to rewrite the forbidden fruit myth to liberate women from negative stereotypes inscribed in mythology, which could otherwise define their lives. The menu of the fruit options that the Goblins are hawking in the opening lines of Christina Rossetti's poem includes all the fruit species employed as motifs in the Pre-Raphaelite paintings I have been discussing – "apples", "oranges", "cherries", "strawberries," and "pomegranates" – as if Rossetti were announcing her intention to exorcize these fruits of their cultural curse.

Works Cited

Barringer, Tim. *Reading the Pre-Raphaelites*. New Haven: Yale UP, 2012. Print.

Bentley, D. M. R. "The Meretricious and the Meritorious in *Goblin-Market*: A Conjecture and Analysis." *The Achievement of Christina Rossetti*. Ed., David A. Kent. Ithaca: Cornell University Press, 1987. Print.

Bullen, J. B. *The Pre-Raphaelite Body: Fear and Desire in Painting, Poetry and Criticism*. New York: Oxford UP, 2005. Print.

Carpenter, Mary Wilson. "'Eat Me, Drink Me, Love Me': The Consumable Female Body in Christina Rossetti's *Goblin Market*." *Victorian Women Poets*. Ed., Tess Cosslett. New York: Longman, 1996. Print.

Garrett, Henrietta. *Wives and Stunners: The Pre-Raphaelites and their Muses*. Basingstoke: Macmillan, 2012. Print.

Gilbert, Sandra and Susan Guber. "The Aesthetics of Renunciation." *Victorian Women Poets*. Ed., Tess Cosslett. New York: Longman, 1996. Print.

Holt, Terrence. "'Men Sell Not Such in Any Town': Exchange in *Goblin Market*." *Victorian Women Poets*. Ed., Tess Cosslett. New York: Longman, 1996. Print.

Mermin, Dorothy. "Heroic Sisterhood in *Goblin Market*." *Victorian Women Poets*. Ed., Tess Cosslett. New York: Longman, 1996. Print.

Pearce, Lynn. *Women/ Image/ Text: Readings in Pre-Raphaelite Art and Literature*. Toronto: University of Toronto Press, 1991. Print.

Polhemus, Robert M. "John Millais's Children: Faith and Erotics: *The Woodsman's Daughter*." *Victorian Literature and the Victorian Visual Imagination*. Ed., Carol T. Christ and John O. Jordan. Berkeley: University of California Press, 1995. 289-312. Print.

Rossetti, Christina. *Poems and Prose*. London: Longman, 1994. Print.

This is Your Brain on Wheat: The Psychology of the Speculator in Frank Norris' *The Pit*

Graham Culbertson

Abstract: This essay argues that Frank Norris's novel *The Pit* should be read as an investigation of the powers of human will and prediction, as against the productive power of the natural world. I show that the novel's protagonist, Curtis Jadwin, seeks to corner the entire world wheat market by outthinking the forces of nature itself. Jadwin's ultimate failure is presented as inevitable because the power of wheat to grow proves to be greater than the power of his mind to anticipate it. As such, Norris's work serves a reminder that despite our illusion of mental control over the markets, the natural word is the ultimate source of productive power.

In the two completed volumes of Frank Norris's trilogy *The Epic of the Wheat, The Octopus: A Story of California* (1901) and *The Pit: A Story of Chicago* (1903), human conflicts play out amidst the industrial production of wheat. Both novels feature human winners, or at least humans who particularly benefit from wheat production, but the ultimate winner in both stories is the wheat crop itself. Norris's aim in both novels is to show that when the wheat crop "wins" – that is, when it overcomes the barriers set in front of it by individual humans, humanity as a whole reaps the benefit. Humankind will do better when the wheat flows freely, and when the individual humans who stand in the way of the wheat are inevitably destroyed. As Norris himself put it in a letter to Isaac F. Marcosson: "The wheat motive is continued the same as in 'The Octopus' – a great and resistless force moving from west to east, from producer to consumer; benevolent and beneficent as long as it is unhampered, but destroying all things and all individuals who attempt to check or divert it" (qtd in McElrath and Jones, xx). But although *The Octopus* has been considered the better and more fully developed of the two novels for the last half-century, certain aspects of the wheat's final victory in that novel leave much to be desired. Donald Pizer states a view that is now quite conventional, that *The Pit*'s love story between Curtis Jadwin and Laura Dearborn is particularly weak and undermines the entire novel. As he puts it: "unlike the love stories of *The Octopus*, [the love story in *The Pit*] seems unrelated to the novel's epic theme of nature's power and benevolence. As a result of this flaw, the epic theme in *The Pit* lacks force, and the work as a whole seems disjointed" (176).

If it is true, as Warren French argues, that "the story of the corner provides only background for understanding the critical education of Laura Dearborn Jadwin" (108), then *The Pit* is largely dismissible. However, there is something very important going on in the novel's depiction of the corner and, in particular, Jadwin's psychology. Jadwin becomes addicted to speculating, like any gambler or alcoholic, and from the beginning, the psychological aspect of the novel has been considered risible. As the contemporary *New York Times* review put it, ultimately the novel merely shows, in needlessly "laborious" fashion, "that speculation in the necessities of life, when conducted on a large scale, is an immoral practice, and very wearing on the nerves of the speculator" (34). Speculation, besides being dangerous, is "very wearing on the nerves of the spectator" – this is the extent to which the anonymous reviewer credits Norris's insight into Jadwin's psychology. In the realm of literary criticism, the template for a similar response was set by Alfred Kazin's cutting remark that the novel "proved no more than that the hold of the pit over Curtis Jadwin was exactly like the hold of drink over the good but erring father in a temperance novel" (78). But this simplistic reading of Jadwin's breakdown, in which speculation is a drug like any other, is insufficient. In *The Pit,* Norris is putting speculation on trial, but speculation of a particular kind. Jadwin's speculation is not mere gambling, but the attempt to bend the productive force of the Earth to his will, and his psychological breakdown is not just the story of a straying husband who has dallied with an addiction, but of a human being who has attempted to become an omnipotent force. The question at hand is, Can any human mind master and overpower the market itself? And it is my contention that *The Pit* does provide a worthy explanation, rooted in psychology, for why such self-aggrandizement is impossible

First, I would like to deal with *The Octopus*, and show why that novel fails to clearly demonstrate that the wheat will always defeat its manipulators. As French puts it, "Norris accomplishes his purpose in *The Pit* without recourse to the mystical trappings and remarkable coincidences of *The Octopus*" (116). In *The Octopus*, S. Behrman, a manipulator of railroads, real estate, industry, and, above all, power, sits in the same position as *The Pit*'s Curtis Jadwin: he is manipulating the movement of the California wheat crop for his own personal profit. But whereas Jadwin is his novel's protagonist, a moderately sympathetic figure who is redeemed through the love of a good woman, S. Behrman is the closest thing that the highly diffuse *Octopus* has to a single villain. Most of the suffering in that novel can be traced to his manipulation of events. If there is one figure that the wheat must overcome to secure its role as a benevolent world force, it is S. Behrman.

And so the wheat does. Near the end of the novel, Behrman goes to inspect his conquest: the physical wheat crop being loaded into a ship headed for the Indian subcontinent. As he watches the wheat flood into a hold of a ship, the "cataract on a sudden increased in volume. He turned about, casting his eyes upward toward the elevator to discover the cause. His foot caught in a coil of rope, and he fell headforemost in the hold" (*Octopus* 1090). Thus is the master manipulator of the novel defeated by the wheat. And Norris leaves no doubt that it was the wheat itself that overcame Behrman: "No human agency seemed to be back of the movement of the wheat. Rather, the grain seemed impelled with a force of its own, a resistless, huge force, eager, vivid, impatient for the sea." (1089-90). The wheat itself is trying to get into the ship, away from the farms of California and into the world market where its life-giving benevolence can be shared with some other corner of the world. Behrman, though he has managed to destroy or suborn all of his human enemies, is for the wheat an inconsequential speck who happens to have impeded it. The movement of the wheat itself is unalterable; as Behrman goes under, Norris concludes the novel's penultimate chapter with an ode to the wheat: "persistent, steady, inevitable" (1093).

Norris concludes the novel's final chapter with another ode to the wheat. This time the subject is not the wheat's greatness but its benevolence: "Through the welter of blood at the irrigating ditch, through the sham charity and shallow philanthropy of famine-relief committees, the great harvest of Los Muertos rolled like a flood from the Sierras to the Himalayas to feed thousands of starving scarecrows on the barren plains of India" (1097). This triumph of capitalist trade means that "[t]he larger view always and through all shams, all wickedness, discovers the Truth that will, in the end, prevail, and all things, surely, inevitably, resistlessly work together for good" (1098). Although humans might think that they need to work to bring goodness to the world, relieving suffering with charity and philanthropy, Norris is revealing that capitalism and the bounty of the Earth will work together to help the worst off. Charity is a mere distraction to the real show; it's the systems of production and trade that will improve the world, and anyone who works as part of the system is working toward an eventual good. Even, it would seem, S. Behrman! But if that is the case, why did S. Behrman have to die? In the novel's logic, since Behrman was manipulating the flow of the wheat to his own advantage, eventually he would have done more harm than good. Norris thus sells Behrman's death as part of the inevitable and resistless benevolence of the wheat, but, in fact, it happened because he tripped. This chance and unpredictable event is no more "determined" by the industrial production of wheat than any other stubbed toe or sprained ankle. This insistence that an obviously chance event lead to an "inevitable" outcome is a feature of both Norris's writing as well as of Zola's, his model.[1]

While Norris's exposition proclaims that the fate of his characters is inevitable and that the power of world-forces is unassailable, his narrative demonstrates a more universally accepted truth: that there is a subtle and unpredictable interaction between forces and people and chance events, and every outcome is the result of the synthesis of many factors. Behrman's death was not inevitable due to plot events that Norris had previously put into play. It was instead a chance event, inserted in the last ten pages of the novel, to get the plot to twist to fit Norris's theory. Norris's Panglossian faith that capitalism and nature's bounty would always pull together to make sure that everyone got fed, regardless of the specific actions of individual magnates, is proven false every time a child starves and a corporation experiences a simul-

[1] In Zola's *L'Assommoir* (1877), a working-class family descends into poverty, alcoholism and death when the breadwinner, the roofer Coupeau, trips and injures himself, despite having been reliably surefooted for the rest of the novel. Norris's *McTeague* (1899) is even more preposterous in its promise of "inevitability;" in that novel, the precipitating event that leads the working-class family into poverty, alcoholism, and murder is the winning of a lottery!

taneous rise in profits. While *The Octopus* is a considerably more thrilling and dramatic novel than *The Pit*, it ultimately fails to prove that wheat is a benevolent force that will overcome those who resist its benevolence. Behrman falls, yes, but it is ludicrous to believe that he "had to."

The Pit, by contrast, suggests real reasons for the fall of its manipulator, reasons which deserve our attention. The Pit itself is the locus of the wheat trade, and for that reason, the second novel does a considerably better job of showing what happens when the benevolent force of the wheat is resisted, when bears and bulls push the market against the flow of the wheat. Bulls, who bet on the market to go up, and bears, who bet on it to go down, are perceived very differently. Walter Benn Michaels, with quotes from Cedric B. Cowing, argues: "The general public [...] is always overwhelmingly bullish, investing for a rise and regarding the great bull traders, in Cowing's words, as 'builders' and 'doers' with 'great dreams of the future.' The bears, on the other hand, were 'the manipulators whom the public feared and resented.'" (73). But Michaels reveals an even deeper reason for the American public to mistrust the bear:

> For not only does he profit from the misfortune of others, not only does he wizard their hard work into the *cause* of their misfortune, he does all this in conjunction with nature! The bear reveals that nature at her most productive and the unproductive speculator – neither growing nor making anything, not even, when he sells short, *owning* anything – are collaborators. The bear's world is a nightmare version of the American dream in which the promise ("Three Harvests more") of unprecedented plenty has turned into the threat ("All greater than you Wish") of overproduction. The land so bountiful no one need work becomes a land whose bounty succeeds somehow in starving instead of feeding its inhabitants. (74)

Michaels repurposes lines from Michael Drayton's "Ode to the Virginian Voyage" as cryptic warnings; American harvests have exceeded anyone's greatest wishes, but this has depressed the market and hurt American incomes. Michaels has identified the central irony of the market: when American farmers are highly productive, when their hard work is transmuted by nature into a bounty of wheat, the farmers actually suffer reduced prices. Certainly, an individual farmer who is very productive in a down year will be prosperous, but when farmers as a whole produce more grain, farmers as a whole receive considerably lower prices. So while the European peasant who was to be the subject of *The Wolf* can buy cheap bread when nature rewards the American farmer, the American farmer wins precisely when he is not rewarded with an enormous harvest.

All of this is amplified by the intervention of the pit, the nexus where American production and European consumption meet. If the market were perfectly efficient, such fluctuations would be relatively minor. But when the harvest is good and prices are low, bears push the prices even lower, at the expense of the farmers and to the benefit of the consumers. When the harvest is poor and prices are high, bulls push the price even higher, at the expense of the consumers and to the benefit of the farmers. When either bears or bulls are operating, more suffering is created than is otherwise necessary. Or this, at least, is Norris's thesis in both *The Octopus* and *The Pit*: that nature's bounty is prevented from reaching a proper equilibrium when a man, or a group of men, intervenes to alter the natural flow of the wheat. Whether that intervention takes place in the form of a conglomerate of ranchers or a railroad trust or a cabal of bears or (most rarely) a Great Bull, the result will be widespread suffering. But as we have already seen, in Norris's fiction, that suffering is always only temporary: eventually the wheat itself will erupt, destroy its man-made bonds, and restore harmony by virtue of its unalterable life-giving essence. It is better, however, not to make nature take that step. Cressler, a stolid trader who eventually chooses to speculate and loses his fortune and his life, explains the system to Laura Dearborn:

> It's like this: If we send the price of wheat down too far, the farmer suffers, the fellow who raises it; if we send it up too far, the poor man in Europe suffers, the fellow who eats it [...] The only way to do so that neither the American farmer nor the European peasant suffers, is to keep wheat at an average, legitimate value. (*Pit* 115).

Bulls and bears, by driving the market illegitimately high or low, ultimately hurt everyone.

As Pizer puts it, speaking of *The Octopus*, "Norris establishes a conflict not between nature and the machine but between accommodation to nature and opposition to it (or, in moral terms, between love and selfishness)" (149). Although humans have built enormous systems of production, transportation, and trading, those systems actually enhance, rather than detract from, nature's power. Kathryn Dolan notes that the "wheat" which Norris is idolizing is actually "a hybrid entity, wheat, an agricultural crop that is neither nature nor culture" but a synthesis of the two (295). That hybrid entity, the combined force of both the natural world and human culture, is in Norris's conception inviolable. As Gina M. Rossetti explains, although Norris is invested in alleviating suffering, he's not actually out to prove that one shouldn't manipulate the market to avoid suffering. He's out to prove that manipulating the market is impossible:

> What the text will show is that speculation is wrong – not because it is manipulative and uses the world's food supply as an exploitative weapon – but because interstate and international commerce are such that there no longer exists a purely American grain of wheat. In other words, wheat cannot be hoarded – as Jadwin desires – to ratchet up prices. Instead, the interconnecting networks come to represent a natural order of things – an "invisible hand" that governs the market benignly. (43)

But although Rossetti is exactly right about both how the markets prevent manipulation in the long-run, I want to focus on the way that this doesn't seem to be true in the novel in the short-run. Certainly, the market is such that Jadwin can't easily hoard wheat, but through the correct manipulation of the futures market, he can ratchet up prices for a time.[2]

When nature and the interconnected market are pulling in a certain direction, the speculator with enough capital and highly accurate information can be enormously profitable. One can be a bear or a bull for a little while, turning a gentle upward or download slope in prices

---

[2] William Cronon explains how the system of Chicago trade actually made speculation possible in the short run: "Chicago's great innovation in the wheat trade had been to simplify the natural diversity of wheat, corn, and other crops so that people could buy and sell them as homogenous abstractions" (132). Arbitrary categories like "No. 1 wheat" versus "No. 2" wheat allowed buyers and sellers to trade in wheat without worrying about its specific quality, while at the same time knowing that wheat "of one grade could not legally be used to fulfill contracts for grain of another." This "artificial partitioning" of the wheat is what made corners possible, since the speculator could use futures contracts to buy a great deal of "No. 1 wheat" before it even existed, and know that no other kind of wheat could be used to fulfill his contract. This allowed for the necessary market distortions for corners. Although Rossetti is right that there is no true American wheat that can be controlled, one can, with enough luck and skill, control the arbitrary abstraction "No. 1 wheat."

into a terrifying rise or plunge. From the beginning, Jadwin's broker Gretry tells Jadwin that a corner is impossible: "They'll slaughter you in cold blood" (173). And Gretry continues to insist that the corner is impossible until the day that Jadwin actually swings the corner. But although Jadwin's corner ends in failure – just like the corner of Joseph Leiter, Norris's real-life model for Jadwin[3] – it certainly didn't have to. Like Leiter, Jadwin could have sold while he could demand his own prices. He chooses not to because he has become gripped by the thrill of control – because his internal psychological process tells him that he will keep winning. Since Norris shares Gretry's views that corners are ultimately futile, it is tempting to regard Gretry as Norris's mouthpiece, a clear-eyed thinker compared to Jadwin's monomaniac. But Gretry is a fool. Although he tells Jadwin that corners are impossible, it is not because he has the wisdom to see that the interconnected systems of nature and humanity will work together to enforce efficiency and alleviate suffering. He simply thinks that the bears will always win. Jadwin, of course, proves him wrong.

So how does Jadwin do this? He does this because he is intellectually bigger and stronger than other men. Unlike *The Octopus*, which featured the physical wheat, the wheat in *The Pit* is "less of a physical and more of psychological presence" (Dolan 306). It's Jadwin's mind and its ability to model and predict the market which makes the corner. Jadwin is particularly good at intuition or unconscious data processing: "Every now and then during the course of his business career, this intuition came to him, this *flair*, this intangible, vague premonition, this presentiment that he must seize Opportunity" (79). Norris goes on to describe "the presence of a new force. It was Luck, the great power, the great goddess," but it is obvious that what Norris is calling "Luck" is actually the moving of the market, the moving which Jadwin's "strange sixth sense" is in touch with.[4] And Jadwin's intui-

---

[3] Norris made a few changes from Leiter's actual corner: "Norris' principle changes were to reduce the length of the corner from over a year to three months, to eliminate the Spanish-American War as a factor, and to subordinate the role of rival speculators in Leiter's collapse" (Pizer 163). Of those changes, the last is the most telling; Norris wanted to depict a battle between man and nature, not a man and a group of men.

[4] As with *McTeague*, the actual events are set into motion by an event of pure chance. When Gretry first tells Jadwin that he should speculate, and Jadwin resists him, Jadwin decides to flip a coin for it. "And as a matter of course the coin fell heads" (80). And Jadwin tells himself that this event was determined: "It will come heads. It could not possibly be anything else. I *know* it will be heads" (80). In his attempt to

tion is unfailingly accurate. His initial speculation comes after a specific tip, but soon he is predicting how the entire wheat market will move.

For this first speculation is a bear speculation. But after getting the taste for speculation, Jadwin decides to pull against the entire market. And he does so because he is processing the data correctly: "Jadwin began to suspect that the wheat crop of his native country, which for so long had been generous, and of excellent quantity, was now to prove – it seemed quite possible – scant and of poor condition." (170). Jadwin bases this intuition on reports from wheat centers in America, from a close look at weather predictions, and from confidential informants in the buying centers all over Europe. His brain has processed all the data and is telling him to buy. This reversal of the market will alleviate the suffering that the bears are causing; although European bread is cheap, "in the United States the poverty of the farmer worked upward through the cogs and wheels of the whole great machine of business" (167-168). But Jadwin senses that this can change: "Jadwin sensed a more rapid, and easier, more untroubled run of life blood. All through the Body of Things, money, the vital fluid seemed to be flowing more easily" (170). The poor wheat crops will synergistically act with this new flow of money, and the result will even out the international market.

The bears resist the new direction Jadwin is taking the market: "In the face of the rumours of a short crop they kept the price inert, weak" (171). At this point, I would ask: Within the novel's internal logic, is the bull market inevitable? According to what we have learned from *The Octopus*, it seems as though it must be. Greater American liquidity and poor wheat crops must drive the prices up; the wheat will accept no long-term manipulation. But there's a possibility that the bears would have kept winning, in the counterfactual situation in which Jadwin plays no role. When Jadwin declares his intention to back a bull market, Gretry tells him, "You're just one man against a gang – a gang of cutthroats" (173-174). Gretry thinks that Jadwin will lose because he is alone, but he wins because he is a superior specimen of humanity. Jadwin, like Frank Cowperwood, the titan of Dreiser's *Trilogy of Desire*, is too cunning and powerful to be defeated by a gang of

render the events of the novel inevitable, Norris simply sweeps the forces of the great goddess Luck in with everything else.

opposing investors.[5] Since Jadwin got into the field of speculation by flipping a coin, what would have happened if the coin flip had gone the other way? One possible answer is that he simply would have bided his time and flipped another coin at a later date, and another, and another, until his latent desire to speculate was affirmed by the chance workings of the universe. But if the coin *is* a good faith effort to let the universe decide, what would have happened, in this particular fictional model of the world, if Jadwin had not gone into the speculation business? If any of the individual bears had disappeared, circumstances are likely to have been more or less the same. But the "Great Bull" is responsible for the corner alone, save for a few men acting as his agents, and with the corner came the corresponding surge in prices, the enrichment of the American farmer and the American market, the impoverishment of the European peasant, and, finally, the bursting of the wheat bubble and the market crash.

The singular nature of Jadwin's corner means that *The Pit* deals with the role of the individual in a highly direct way. Gretry tells Jadwin, "you are either Napoleonic-or-or a colossal idiot" (174). Although Jadwin responds that he's "just using a little common sense," Grety's metaphors ring true. The wheat crop in America was going to be bad whether or not Jadwin won his coin toss, and the American investment climate was likewise going to be good, but I would like to suggest that Jadwin's corner would not have happened without Jadwin. If there had been others well suited to be the Great Bull, there would have been a few Great Bulls. To continue with Gretry's Napoleonic metaphor, let us borrow a question from William James's "The Importance of Individuals": "Some organizing genius must in the nature of things have emerged from the French revolution; but what

[5] Although Cowperwood is probably the closest analogue to Jadwin in the literature of this period, there are vast gulfs between them. What they both share is a superiority to those around them in force of will, intellect, and predictive powers. This allows both of them to foresee where the market is going and to organize business ventures to take advantage of their foresight. But Cowperwood is an *Ubermensch* in almost every way. He discards conventional sexual morality, thrives on business and sexual conquests, masters the realms of art and culture, and blatantly flouts all ethical, political, and legal systems set up to govern his behavior. Jadwin, by contrast, is presented as essentially bourgeois in his sexual relationships (a marriage which is only threatened by his work), his tastes in art and literature (Howells is his favorite novelist), and his business dealings. Ultimately, the greatest difference is that Norris regards Jadwin as a pawn in the face of the cosmic force of the wheat, while Dreiser's Cowperwood truly embodies certain life-forces.

Frenchman will affirm it to have been an accident of no consequence that he should have had the supernumerary idiosyncrasies of a Bonaparte?" (651). In the novel, Jadwin worries, before he has begun his career as the Great Bull, that another will step in and take that position. "Was this upheaval a revolution that called aloud for its Napoleon? Would another, not himself, at last, seeing where so many shut their eyes, step into the place of high command?" (229). But even if, as Norris and James both seem to agree, another Napoleonic figure would have emerged in these revolutionary conditions, the fact remains that the exact dimensions of that particular figure are of tremendous importance.

To put it a different way, I take it to be absolutely crucial that Jadwin, as an individual, behaved exactly as he did. The ranchers and the railroads of *The Octopus* and the bears of *The Pit* behaved selfishly, opposing the combined will of the wheat and will of the market, and they were punished for it. However, the railroad, like the ranchers and the bears, seems to have behaved more or less autonomously, and not to have been directed by a single individual. This collective, force-driven behavior is explained by Shelgrim, the railroad president, in the most famous passage of *The Octopus*: "Control the Road! Can I stop it? I can go into bankruptcy if you like. But otherwise, if I run my road, as a business proposition, I can do nothing. I can *not* control it. It is a force born out of certain conditions, and I – no man – can stop it." (396) In Shelgrim's description, the railroad is beyond human agency with only a single fail-safe: bankruptcy. Shelgrim can choose to charge less than the market will bear, and the result will be the demise of the railroad. In Shelgrim's account, these are the only two options: either the railroad operates with complete disregard to the wants and desires of humanity, to anything except to market conditions, or the railroad ceases to exist.

But not so for Jadwin's corner. The Great Bull alone decides whether or not to attempt a corner, whether or not to give up the corner, and whether or not to sell and reap the profits of his corner at any given time. This means that the corner is a special circumstance: for a certain period of time, a single individual and his psychology can affect the entire market. Jadwin can and does drive the market exactly where he wants it to go by sheer, individual will – until he reaches his inevitable failure and the corner collapses. The novel, once again via Gretry, provides a conventional explanation for why the corner collapses and Jadwin is ruined. Gretry's argument that one man cannot stand against the bears is proven false; Jadwin ruins the bears. But human psychology cannot defeat the larger system of the wheat. Both the novel and Jadwin repeatedly embrace this theory: Jadwin is not truly in control, but a pawn. It is not his will that drives the market, but the force of the wheat. At first, Jadwin believes himself to be master of the situation: "Now he had discovered that there were in him powers, capabilities, and a breadth of grasp hitherto unsuspected. He could control the Chicago wheat market; and the man who could do that might well call himself 'great,' without presumption" (*Pit* 247). This passage emphasizes his powers, his capabilities, his triumph. But just a few days or weeks later, Jadwin no longer feels full of force but, rather, his "tired brain flagged and drooped" (248). The power driving the corner is inside Jadwin's mind, but it isn't Jadwin himself. His brain echoes, over and over again, with the mantra "Wheat-wheat-wheat, wheat-wheat-wheat!" (248). And he, like Shelgrim, disavows the actions of the system he nominally controls. He tells his wife:

> You think I am wilfully doing this! You don't know, you haven't a guess. I corner the wheat! Great heavens, it is the wheat that has cornered me. The corner made itself. I happened to stand between two sets of circumstances, and they made me do what I've done. I couldn't get out of it now, with all the good will in the world. (249)

A mere two pages ago, Jadwin's internal monologue tells him that it is his capabilities and powers that have made this corner possible, but now, in this dialogue with his wife, he is disavowing all responsibility for the corner – the wheat made him do it. Which of these perspectives on his role should we embrace? From what we know of *The Octopus*, it seems likely that Jadwin is merely a pawn.

Jadwin is going through a metamorphosis in his self-image: from an individual agent, to the pawn of outside forces, and finally to the human avatar and manifestation of those forces. In the first stage, he believed that he, acting alone, was going to swing the entire corner of his own powerful will. In the second stage, he has disavowed all agency and ascribed the entire corner to circumstances, to the wheat, to the market, to anything except for Curtis Jadwin. After he has pulled off his corner, he has decided that there is neither inside nor outside, but that Curtis Jadwin and market forces are one and the same. When

Gretry, predicting the future correctly for the first time since Jadwin's corner began,[6] tells Jadwin to sell, he puts it in stark terms: "'Great Scott J., you're fighting against the earth itself'" (304). Jadwin responds: "'Well, we'll fight it then. What do I own all these newspapers and trade journals for? We'll begin sending out reports tomorrow that'll discourage any big wheat planting.'" Jadwin has mistakenly come to believe that he *is* the earth and its wheat crop, that whatever he commands, nature itself will fulfill. As William B. Dillingham puts it, "Jadwin is a level-headed self-made man who is caught up in a foolish attempt to make himself a god by controlling a great natural force" (96). Because he has become deranged by the power of the wheat, Jadwin has gone from level-headed to self-idolatrous. In my reading, he can't tell the difference between himself and the wheat because it has invaded his psychology and essentially remade his mind in its own image. The word wheat echoes through his consciousness endlessly, and he has the "permanent" feeling of an "iron clamp about his head," the manifestation of the wheat's control (305).

But Jadwin is mistaken about the nature of the transformation. The wheat has taken over Curtis Jadwin's mind, but Jadwin's mind does not control the wheat. The wheat, in fact, has worked to destroy him, just as it destroyed Behrman. And it doesn't need silly tricks to do it. Jadwin has extended his essence so that it is tightly intertwined with a poor wheat crop. All the wheat has to do to destroy him is grow. And grow it does:

> It was the Wheat, the Wheat! It was on the move again. From the farms of Illinois and Iowa, from the ranches of Kansas and Nebraska, from all the reaches of the Middle West, the Wheat, like a tidal wave, was rising, rising. Almighty, blood-brother to the earthquake, coeval with the volcano and the whirlwind, that gigantic world-force, that colossal billow, Nourisher of the Nations, was swelling and advancing. (326-327)

---

[6] Gretry is a perfect hedgehog; he knows one big thing: that the market will always have downswings. If the market happens to be in or near a downswing, he will appear wise. If not, he will appear foolish. In this case, he is right because Jadwin has driven the price of wheat to absurd heights, but when it was at absurd lows he was giving the same advice.

Although it had obviously used tools such as Jadwin and the farmers along the way, these tools are ultimately irrelevant: "What had they to do with it? Why the Wheat had grown itself; demand and supply, these were the two great laws the Wheat obeyed." (327). Jadwin's downfall was inevitable all along, because he was opposing an ancient and all-powerful force. "Almost blasphemous in his effrontery, he had tampered with these laws, and had roused a Titan." In Dreiser's novel of Chicago business, *The Titan*, it is the market-manipulator who becomes an irresistible force, but Norris values the strength of humanity at a much lower level. At the conclusion of *The Pit*, the wheat is once again dominant, and it seems that the role of the individual human – even the Napoleonic man – is negligible.

We can go beyond this conclusion when we realize that *The Pit*'s true contribution to the literature of markets and industrial production is Jadwin's psychology. If Jadwin had acted differently at a single precise moment, he wouldn't have failed. If he had sold when Gretry had told him that the wheat had turned against him, when it was worth $1.50 and only he had access to any wheat, he could have made an enormous fortune. Jadwin could then have played the bear market for a while, harming farmers and helping wheat consumers, before getting out again at the right moment. With the vast wealth gained from his bull corner, he could even weather a number of missteps; as Gretry explains to Jadwin when he tries to push the corner a bit too far, it took "every cent of our ready money to support this July market" (323). And if he had closed his corner instead of pushing it too far, he would have had enough money to prop up any future market. But he didn't, because the wheat had convinced him that he was irresistible. The reason that Jadwin is a blasphemer against titanic forces, rather than a new, human manifestation of them that he believed himself to be, is that his mind fails. Norris describes the effects on Jadwin's psychology when the market crashes:

> And then, under the stress and violence of the hour, something snapped in his brain. The murk behind his eyes had been suddenly pierced by a white flash. The strange qualms and tiny nervous paroxysms of the last few months all at once culminated in some indefinite, indefinable crisis, and the wheels and cogs of all activities save one lapsed away and ceased. Only one function of the complicated machine persisted; but it moved with a rapidity of vibration that seemed to be

> tearing the tissues of being to shreds, while its rhythm beat out the old and terrible cadence:
>
> "Wheat-wheat-wheat, wheat-wheat-wheat." (343).

By the end of the novel, it has become a common opinion that it is the wheat itself that both made Jadwin's corner and destroyed it. Laura quotes Jadwin's claim that "'[t]he wheat cornered me, not I the wheat'" and Crookes, the bear leader, is contemptuous of his fellow bears, explaining "*They* didn't do it. It was the wheat itself that beat him" (368, 347). Jadwin has been winning his battles because his mind has managed to model the world-wide wheat situation. His mind is, as in Hobbes's account, a set of wheels and cogs that, in this case, are busy calculating the wheat's next move. In order to keep up with the wheat, Jadwin has been experiencing progressive decay as more and more of his processing power is taken up anticipating the movement of the wheat. This accounts for his disengagement from his wife, his feeling of a dull ache in his mind, and the "strange qualms and tiny nervous paroxysms" that have been afflicting him. The physical wheat has been transformed by the pit into an abstract, disembodied force, and that force has infiltrated Jadwin's brain. Once there, it becomes embodied once again in Jadwin's physical being; the pit is not the tool Jadwin uses to become the wheat, but rather the tool that the wheat uses to take over Jadwin's body. As Katherine Biers explains: "It is not so much the psychological but the physiological effects of" the wheat on Jadwin that Norris depicts at the end of the novel (524). When the wheat suddenly surges the other way, springing up all over the world instead of receding, it is not Jadwin's external machinery of trade and information-gathering that fails him. It is his internal circuitry that snaps, as Jadwin's "immersion in the trading pit produces physical effects on his brain and nervous system" (Biers 524).

As both Christophe den Tandt and David A. Zimmerman have remarked, the particular method by which Jadwin's body has been hijacked is mesmerism. As den Tandt puts it, "the mesmerizing energies of speculation stir in Curtis Jadwin the fantasy that the economy of the city can be captured and unified by means of brutal strength and willpower" (90). Zimmerman expands this analysis, explaining that not only has Jadwin been mesmerized by the power of speculation, but Jadwin himself is also "a mesmerist" (135). The wheat crop has been, for a time, mesmerized by the speculator, a metaphor which Zimmerman explains to have been so common as to be a cliché. But in this system of mutual mesmerism, there can be only one eventual outcome: the triumph of the natural force. In trying to capture the market, Jadwin has come up against the sublime, "the terror, the inability of the single mind to master and recuperate such sensory excess" (127). Representing the market in a novel such as *The Pit* is a problem of language, but Jadwin's attempt to control the market is a problem of self. Although Jadwin thinks that he has become the market by mesmerizing it, the flow of force is much stronger in the other direction. "Jadwin tries to appropriate and contain an excess – semiotic, feminine, natural, mechanical, even religious – that defies and overwhelms his powers of representation" (142). This sublime excess overwhelms his mind, because it is not up to the task of modeling the entire market and all of nature. Jadwin's brain has grown tight, and eventually snaps, under the pressure of attempting to contain too much.

This leads us to two alternate conclusions about humankind's ability to control global markets driven by Mother Nature. The first conclusion, and the one which Norris clearly intends, is that such control is impossible. In *The Octopus* we learned that external systems built to help us control nature will work with her and take on a life of their own; the wheat uses the railroad to crush human opposition, rather than the other way around. In *The Pit* we learn that the internal, psychological systems we build to help us control nature will ultimately be inadequate to the task; Jadwin's mind is driven by the wheat until the wheat no longer needs it, and then it collapses. An alternate conclusion, hearkening back to William James, is that Jadwin was merely not the right man for the job. A stronger and more nimble thinker, one less likely to get on a single track and better able to switch gears, might well have been able to thrive at the moment of crisis, flow with the wheat in its new direction, and become the indestructible kind of titan that I outlined earlier. Whether or not we think that Norris has proven that the forces of the wheat will always overwhelm the human actor depends on what we think about Jadwin's unique nature. Before he is taken over by the wheat, Jadwin is a highly talented man, a forceful organizing genius blessed with tremendous intuition.[7] A range

[7] Jadwin has another tremendous advantage, particular to his historical moment. In his book *American Nervousness: 1903*, Tom Lutz identifies nervousness and neurasthenia as ubiquitous phenomena in American culture. Even within that moment, *The Pit* stands out as a "neurasthenic tale" in which all of the major characters are nervous

of fellow characters, from his enemy Crookes to his wife Laura, praise him as a particularly strong-willed and far-seeing man. If he is in fact the best that humanity has to offer, then it is hard to imagine any human outwitting the forces of the wheat. But if it is possible for a figure of even greater force and intellect, a Frank Cowperwood, to have stood in his place, perhaps a human titan could have emerged.

But even Cowperwood busts in each of his first two novels. Although he and his descendants, such as Welles's Charles Foster Kane and DeLillo's Eric Packer, seem to be more forceful and agile, more imbued with a Nietzschean superiority, than the stolid Curtis Jadwin, they all run up against the moment where their ability to predict and will the movement of the market fails them. By creating a man of business whose ability to prosper fails at the moment that that he becomes psychologically locked into a single path, Norris has created a durable archetype of the American business titan. And by repeatedly emphasizing that it is always nature lying behind that market and thus overwhelming human psychology, Norris's novel serves as a reminder that that the market and the mind are not free-floating systems always only interacting with one another, but are inevitably tied back into the forces of the surrounding world. Nature, whether via brute force or psychological mesmerism, will always win its battles with human will. If Norris's novel hasn't managed to turn this lesson into an ultimate and absolute truth, as he wished, it has certainly stood the test of time as a cautionary tale.

Works Cited

Biers, Katherine. "Frank Norris's Common Sense: Finance as Theatre in *The Pit*." *Textual Practice* 25 (2011): 513-541. Web. 2 July 2013.

Cronon, William. *Nature's Metropolis: Chicago and the Great West.* New York: W.W. Norton and Company, 1991. Print.

Den Tandt, Christophe. *The Urban Sublime in American Literary Naturalism.* Urbana: University of Illinois Press, 1998. Print.

---

(147). But although Jadwin has the ultimate breakdown, which Norris makes "the most detailed literary representation of a neurasthenic crisis," at the beginning of the novel Jadwin seems exceptionally safe from nervousness, even calming Laura by his stolid presence (Lutz 156, 149). Jadwin should have been the only character safe from nervousness, but even his defenses are no match for the wheat.

Dillingham, William B. *Frank Norris: Instinct and Art*. Boston: Houghton Mifflin Company, 1969. Print.

Dolan, Kathryn Cornell. "A 'Mighty World-Force': Wheat as Natural Corrective in Norris." *ISLE* 19 (2012): 295-316. Web. 2 July 2013.

French, Warren. *Frank Norris*. New York: Twayne Publishers, 1962. Print.

James, William. "The Importance of Individuals." *Writings 1878-1899*. New York: Library of America, 1992. Print.

Kazin, Alfred. *On Native Grounds*. Abridged Edition. Garden City: Anchor Books, 1956. Print.

Lutz, Tom. *American Nervousness, 1903: An Anecdotal History*. Ithaca: Cornell University Press, 1991. Print.

Michaels, Walter Benn. *The Gold Standard and the Logic of Naturalism*. Berkeley: University of California Press, 1987. Print.

Norris, Frank. *The Octopus*. 1901. *Novels and Essays*. New York: Library of America, 1986. Print.

—. *The Pit*. Ed. Joseph R. McElrath, Jr. and Gwendolyn Jones. New York: Penguin Books, 1994. Print.

Rossetti, Gina M. "Turning the Corner: Romance as Economic Critique in Norris's Trilogy of Wheat and Du Bois's *The Quest of the Silver Fleece*." *Studies in American Naturalism* 7.1 (2012): 39-49. Print.

Pizer, Donald. *The Novels of Frank Norris*. Bloomington: Indiana University Press, 1966. Print.

Zimmerman, David A. *Panic!: Markets, Crises, and Crowds in American Fiction*. Chapel Hill: University of North Carolina Press, 2006. Print.

"*The Pit:* A Dispassionate Examination of Frank Norris's Posthumous Novel." Rev. of *The Pit*, by Frank Norris. *New York Times: Saturday Review of Books and Art*. 31 January 1903: 66. Reprinted in *Critical Essays on Frank Norris*. Ed. Don Graham. Boston: G.K. Hall & Co., 1980. Print.

Refusing Form: A Reading of Art, Americanism, and Feminism through Plant Imagery in Susan Glaspell's *The Verge*

Stacey Artman

Abstract: The protagonist of Susan Glaspell's *The Verge* represents some of the tensions that Americans, and particularly female artists in America, experienced during the post-war era. For Glaspell, *The Verge* and its protagonist provide a literary space to test new forms of art, womanhood, and Americanism. This play may be one of Glaspell's most complex theatrical productions, and this essay employs Conceptual Blend Theory to analyze how the ingenuity of the play lies in Glaspell's use of metaphor to blend Claire, her plants, and the ideals she embodies into something "other," something on the verge of new.

After the violence and technological advances of World War I, Americans were rethinking what was within the verge of human possibility. In an era of uncertainty, rapid transformation, and controversy in nearly all aspects of American life, Susan Glaspell's sensitivity in conveying the sentiment of a nation on the verge of transformation appear in both the content and form of one of her first full-length plays, *The Verge*. Claire, the protagonist, represents some of the tensions that Americans, and particularly female artists in America, experienced during these times of transition. Claire evades definition and, in doing so, provides a literary representation of the conflicts within both the individual and collective experiences of America in the early twentieth century. For Glaspell, *The Verge* and its protagonist provide a literary space to test new forms of art, womanhood, and Americanism.

Although the play received mixed reviews, Kenneth Macgowan noted that "for the tiny audience of the keen, the sensitive, the genuinely philosophical, here is the most remarkable dramatic document that they have ever come across" (qtd. in Gainor 165). Ellen J. Gainor wrote that critiques of the play all had in common the "writers' attempts to provide a totalizing explanation of the play, to bring to it a coherence and sense of containment that the play itself continues to evade" (143). Glaspell evades such containment by constructing "her drama around transitions and contrasts: between characters' beliefs and actions, botanical forms, and states of mind, among others." Ruth Hale, the *New York Times* reviewer of the 1921 New York debut, wrote, "If the surface of life changes by a hair's breadth, she [Glaspell] not only knows it, but can convey it in words" (Nelligan 99). This play may be one of Glaspell's most complex theatrical productions, and the true brilliance of it lies in the ingenuity of Glaspell's use of metaphor to blend Claire, her plants, and the ideals she embodies into something "other," something on the verge of new.

Karen Malpede calls Claire "a revolutionary character in a revolutionary play written by a revolutionary woman" (124). Claire Archer is a bold and privileged woman who is living in New England after the Great War. She uses her greenhouse as a laboratory in which she experiments with hybrid plants to create new and undefined forms of life. Claire's language throughout the play is mostly abstract and even poetic; she expresses herself and her work almost exclusively through abstract philosophical musings and metaphor. She is a complex and misunderstood character that Glaspell uses to comment on the artistic and sociopolitical issues of her time. Glaspell's commentaries, like Claire's character, are neither definitive nor complete. Tom, the character who best understands Claire, explains that she "isn't hardened into one of those forms she talks about. She's too – aware. Always pulled toward what could be" (Glaspell 40). Like Claire's Breath of Life plant, and Claire herself, *The Verge* cannot be contained as a coherent form with a definitive purpose. It is in a position to define the contours of a new era and to transition to new literary forms and social constructs.

The play's historical context provides a framework for investigating how Glaspell uses metaphor to create a drama that is as complex, fluid, and elusive as the experience of the female creative self during the modern postwar era. Glaspell uses metaphor not only as a literary device, but as an instrument for understanding the complexities and nuances of the lived experience of the early twentieth century without attempting to contain them within one coherent description. The plants within the drama become a metaphor for making sense of the complexities of being an individual, a woman, an artist – and all three simultaneously. Although there was no longer a physical 'frontier' in

America, the language and ideology forged by its history had not disappeared from American thought by the 1910s-1920s. During Glaspell's time, the frontier was a sociopolitical and creative one. Like Claire's Breath of Life, Glaspell's *The Verge* is "quite new in form. It – says something about form" (Glaspell 69). Rytch Barber claims that "Susan Glaspell is undeniably the pioneer of this vanguard [expressionism]" (94) and that "Glaspell's use of expressionist dramaturgy in *The Verge* is [...] not simply stylistic experimentation, but a well-calculated effort to find a whole new theatrical language with which to articulate social critique and express a feminist view of American society" (102). Although critics and literary analysts attempt to define Glaspell's work as part of American expressionism, Glaspell resists such containment to a singular form. While she employs expressionist techniques, she combines them with a realistic plot, and also questions the value of expressionism throughout the dialogue of the play. Elizabeth attempts to define Claire's work by describing it as part of the expressionist movement of the time. She explains, "we should each do some expressive thing – you know what I mean? And that this is the keynote of the age. Of course, one's own kind of thing. Like mother – growing flowers" (Glaspell 44). Here, Elizabeth attempts to contain Claire's work within a socially acceptable form, the "keynote" of the age. Tom also supports expressionism as an acceptable mode for understanding Claire's work. He explains to Claire that "things may be freed by expression. Come from the unrealized into the fabric of life" (57). Claire responds, "Yes, but why does the fabric of life have to – freeze into its pattern?" (57). Although she recognizes the value of expressionism, she rejects any type of form. Just as Breath of Life will "a thousand years from now" be "but a form too long repeated" (69), expressionism loses its value when it becomes a repeated, widely accepted form.

This question of a widely accepted form, or code, illustrates questions of Americanism at Glaspell's time. The 1915 article "What Is Americanism?" concedes that "if there be such a thing as Americanism it is not composed into a widely accepted code" (482). The authors' admission that no unified, "widely accepted" definition exists illuminates the historical context of the play's creation. Claire's resistance to being contained creates a correlation between the character and the American people she represents; both resist a unified form and exist within a state of fluidity, on the verge of something yet undefined.

It is the play's resistance to form that opens it to multiple interpretations and methods of analysis. Cognitive literary studies offer a unique framework for understanding Glaspell's play by using Claire's plants as a guiding metaphor throughout the text. This framework accounts for the artistic, social, and personal influences that informed Glaspell's writing of *The Verge*. It also allows for multiple and diverse interpretations, while unifying the overarching themes of the text. George Lakoff and Mark Johnson's *Metaphors We Live By*, first published in 1980, is one of the fundamental texts of cognitive literary studies. The book brought to light the fact that "How we think metaphorically matters" (243) and demonstrated that metaphors are not simply literary devices used for poetic flourish. Rather, they are conceptual tools for making sense of the world. Gilles Fauconnier and Mark Turner draw from Lakoff and Johnson's work and expand upon the study of conceptual metaphor in their book *The Way We Think: Conceptual Blending and the Mind's Hidden Complexities*. The authors introduce what has come to be known as Conceptual Blend Theory. This theory utilizes social, scientific, and artistic studies to explore how humans make meaning of their world and how concepts are formed and manipulated. The authors explain that their theory is an attempt to break away from strictly formal approaches without neglecting them altogether. The authors' analysis of human thought is founded on "what the form approaches have assumed as given: the operations of identity, integration, and imagination" (6). This approach to understanding metaphor and meaning is particularly suited to studies of *The Verge* because, like the play, cognitive literary studies evades formal approaches and is founded upon an interdisciplinary blend, a hybridization, of knowledge. As a theoretical framework, it draws from both scientific and artistic endeavors to provide a holistic approach to understanding "the way we think."

The authors describe cognition as relying upon mental spaces, "small conceptual packets constructed as we think and talk, for purposes of local understanding and action" that "are connected to long-term schematic knowledge called 'frames'... and to long-term specific knowledge" (Fauconnier & Turner 40). There are several types of mental spaces. Input spaces are created by external stimuli; they contain all of the information an individual has about a single concept or event. Generic spaces exist in an individual's consciousness from lived experience; these map onto inputs and contain what inputs have in common with one another (41). The blend, or blended space, is formed from two or more input spaces, in addition to the generic spaces that help to create connections between them (42). This blended space contains elements from each of the input and generic spaces, but also "develops emergent structure that is not in the inputs." In the blended space, connections are available that do not exist within the separate inputs. They are created by the multiple connections that are made simultaneously between input spaces, generic spaces, and frames. Once all of these elements become part of the blend, it is integrated (43) and an individual can "run the blend" in a process called elaboration, the process of utilizing the blended space to achieve understanding, or create meaning (44).

In the case of *The Verge*, there are input spaces created by Glaspell's cultural context, her personal experiences, and the influences of the artistic community. The three primary cultural inputs are Americanism, modernism, and feminism. Elements of these inputs map onto the plants in Claire's greenhouse, and the plants map onto Claire. In turn, both Claire and the plants enter a blended space, eventually developing a larger "megablend" (Fauconnier & Turner 151-153). Claire represents Glaspell's creative self, so Glaspell's life, values, and experiences are also added to the blend so that the megablend encompasses the cultural and artistic influences, Claire, the plants, and the author. Thus, the more that is revealed about Claire and her plants, the more we can understand Glaspell's experience as a female artist in America after the Great War. These relationships are not direct correlations that can easily be explained in terms of x represents y, or in terms of tenor and vehicle, as may sometimes be expected of literary metaphors. Instead, all of the inputs bring elements to the blended space and within the emergent structure of the blend new and unique meanings form from their integration. In keeping with *The Verge*'s refusal of form, the plants function as metaphors, but they do not provide a direct correlation to any one idea or element of the play. Instead, they enter into a blended space where the plants act as the vehicle for multiple tenors, which often overlap and converge with one another. In the language of the play, the blend has "reminiscence" (Glaspell 32) from the input spaces, but also achieves "otherness" (33) through its unique emergent structure and through elaboration.

Glaspell's choice of plants as the focus of Claire's experimentation is particularly suited to the themes of the play because plants have unique characteristics that, individually and combined, lend themselves to a dynamic and elusive use of metaphor. They are aesthetic, experimental, often uprooted and hybridized, and they exist as part of an ecosystem. In addition, they contain both male and female reproductive organs. All of these characteristics offer elements of meaning to the themes and plot of the play. Concepts and structures from each of the inputs influence the blended space and its emergent structure, then elements of that conceptual blend and its emergent structure are mapped back onto the input spaces to give them nuanced meaning within the context of the play.

Claire continuously associates herself with plants. In doing so, she becomes like them in a type of reverse-personification. The two merge so closely that woman and plant become as one literary entity. Claire is even called the "flower of New England" by her husband, Harry (Glaspell 32). Women are often associated with flowers and this phrase comes as no surprise from the voice of Claire's conventional husband. By referring to her as a "flower," Harry also reifies the close connection between Claire and her plants; she is both gardener and flower. Throughout the play, Claire is creating a new form, attempting to redefine the word *flower* in an attempt to symbolically redefine "flower of New England." Tom explains that "If she can do it with plants, perhaps she won't have to do it with herself" (40). Although Tom understands Claire more than the other characters do, he too attempts to define her using the traditional flower metaphor. He refers to her as the "brave flower of all our knowing" (60), assigning to the metaphor more power and intelligence than the traditional flower metaphor that depicts women as purely aesthetic and fragile. Still, even Tom is unable to completely redefine Claire. She later attributes to the existence of Breath of Life the ability "To make a person new" (69). Although Claire is musing about new forms of plants that can come from "the madness" that gave birth to Breath of Life, she quickly turns from discussing plants to discussing people, demonstrating the extent to which the two have become integrated conceptually for Claire. Claire's relationship to her plants demonstrates the conceptual nature of metaphor as it is described by Lakoff and Johnson. Claire is metaphorically called a flower. In many ways, she has internalized this metaphor and identifies herself as a plant, creating a conceptual integration that blends her and the plants. If she can redefine flower and turn a flower into "what hasn't been" (32), then she can also see herself becoming "what hasn't been." She can be revolutionary, like her flowers.

This human-plant blend is made particularly apparent when Tom refers to Breath of Life as the "flower [Claire] breathed to life" (Glaspell 57). This phrasing alludes to the biblical creation story in which the Creator breathes life into the first human beings, the first of their kind, thereby creating a completely new form. This cultural allusion becomes part of the conceptual integration – the blend – and associates Claire's work with that of the Creator and the plants with human beings. The humans of the creation myth were "made in the image" of God, but they were not God. Like Breath of Life, the humans of the creation myth were the first of their kind, but they retained a "reminiscence" of the deity that had come before them. This integration of narratives allows for the creation myth to map onto the original input of the plants and Claire. As Claire becomes more like her plants, the men become increasingly fearful of the plants and what they might mean for society. Through elaboration of the blend, Claire is perceived as both the creator and created, the experimenter and the experiment, and therefore may have the power to create her own identity. At first, Claire's husband, Harry, claims that Claire's experiments are "not important" (29). As Claire's Breath of Life plant is on the verge of blooming, however, Harry's disdain for Claire's work increases. The closer she is to success, to imitating the work of the biblical Creator by "creating" (33) new forms, the more empowered she becomes and the more threatened Harry is by her work, which he eventually refers to as "unsettling for a woman" (33). Although Dick is a more progressive character than Harry, he too finds Claire's experiments to be unsettling because he believes the experimentation to be dangerous. He ponders, "Plants are queer. Perhaps it's safer to do it with pencil (regards TOM) – or with pure thought. Things that grow in the earth –" (42). Tom finishes his thought by stating, "I suppose because we grew in the earth." Again, human beings are associated with plants and the men are unsettled by Claire's manipulation of the plants' forms. This fear of experimentation, and especially experimentation by women, provides a commentary on the social tensions of the turn of the century, when new forms of art were threatening tradition and new forms of womanhood were threatening traditional gender roles.

Claire's plants are artistic, scientific, and social experiments. She and the other characters continuously refer to the aesthetics of the plants, and Claire often speaks of them in abstraction, as though they are pieces of art. Glaspell connects plant life to artistic endeavors immediately in the first act of the play. She describes the frost that has "made patterns on the glass as if – as Plato would have it – the patterns inherent in abstract nature and behind all life had to come out, not only in the creative heat within, but in the creative cold on the other side of the glass" (Glaspell 25). From the onset of the play, nature is given a creative and artistic nature. Yet, there is a very calculated manner with which Claire tends to her plants. Anthony explains that Claire maintains a precise temperature of seventy-three in the greenhouse, and Glaspell describes the greenhouse as a laboratory, suggesting that Claire's horticultural work is as much a scientific experiment as it is an artistic one. It is both inspired (as suggested by the name Breath of Life, a hint toward the etymology of inspire) and calculated. Claire's horticultural projects closely mirror Glaspell's creative endeavors; they are both artistic and experimental. They draw from what has been, but exist in a liminal stance between existing forms and something new. Both women's endeavors are rejected by the traditional voices of their time; Claire's experimentation is deemed as "unsettling" (33) and Glaspell's *The Verge* received many negative reviews. Glaspell's art, like Claire's actual soil, is the ground within which her revolutionary experiments are rooted.

At the same time that Claire is revolutionary, she represents two of the longest standing ideals of American society: progress and experimentation. In the article "What Is Americanism?," the authors claim that "The majority of educated Americans take it for granted that civilization is at present rudimentary, and that it is to develop indefinitely" (434). At a time when World War I and technological advances raised many questions about the stability of America and about how to reconcile the country's past with its ever-pressing desire for progress, Glaspell's creation of a female protagonist (in itself a marker of social and artistic progress) who is obsessed with the idea of progress – though she does not know to what end – reflects the sentiments of the American collective at the time. Claire's inability to define her goals, or even what she considers progress, serves several purposes within the drama. The question of objectives is most apparent in Claire's dialogue with her daughter, Elizabeth, a "cultivated" New England woman (Glaspell 43). In Claire's understanding, Elizabeth is "finished" (42). This description of her daughter is a bit of word play on the idea of "finishing school," and it is a judgment on her daughter. Claire rejects anything that is finished or definitive because it has "gone dead in the form" (33) and no longer verges on the possibility of being "outside." Elizabeth insists that Claire's plants must be "better" (45) and that her mother must work toward a specific goal. She suggests that "it is your [Claire's] splendid heritage that gives you this impulse to do a beautiful thing for the race [...] you are doing in your way what the great teachers and preachers behind you did in theirs." Elizabeth attempts to fit her mother's work into a tradition, a form that is socially acceptable. Claire adamantly denies that there must be a definitive end result, claiming, "They may be new. I don't give a damn whether they're better." Elizabeth, like Harry, condemns this action. She first attempts to justify her mother's work by stating that it is being done "to add to the wealth of the world" (45), but when her mother rejects the idea, Elizabeth eventually condemns the work, stating, "Unless you do it to make them better – to do it just to do it – that doesn't seem right to me" (47). It is in exchanges such as this dialogue that the blend of Glaspell's art and Claire's horticulture becomes apparent. There must be a set of criteria, a form, against which to judge something if one is to deem it "better." It is exactly these criteria that Glaspell and Claire are attempting to escape.

Elizabeth, like the Edge Vine, "isn't – over the edge" and joins in the male figures' condemnation of Claire's experiments (Glaspell 47). Claire rejects Elizabeth and identifies her as the reason why Claire must "leave." Elizabeth, a woman who is conventional, definitive, and happy to remain that way, is "there to hold the door shut." She is as much an impediment to the progress of art, science, and society as the men who condemn Claire's experimentation. Claire is confined as much by women who refuse to experiment as she is by the men who are unsettled by her actions. This scene provides another example of how the plant has become a conceptual metaphor for Claire and guides her decision making. As Claire becomes frustrated with her daughter's conformity, which is made evident by Elizabeth's repetition of the phrase "all the girls" (46), Claire becomes increasingly enraged with the Edge Vine's failure. Eventually she tears it from the ground with her bare hands, explaining, "It isn't – over the edge. It's running, back to – 'all the girls'" (47). For Claire, the Edge Vine and Elizabeth have become one entity, just as Claire and Breath of Life have become one.

Claire condemns the Edge Vine for "going back home" (Glaspell 29), but later states, "We need the haunting beauty from the life we've left" (57), expressing her desire to maintain some elements of the past. She is trapped between a desire to explore new frontiers and the "haunting beauty" (57) of what has come before her. Claire's desire to break from tradition and find "what hasn't been" (32) is rather ironic in light of her ancestry. The very desire to explore, experiment, and break out aligns with the tradition of pioneering passed down through her family history. In *The Verge*, the theme of pioneering encompasses the concepts of forging something new and migrating to a new environment. In Ben-Zvi's biography of Glaspell, she explains, "Pioneer [...] defines the direction of Glaspell's own life and the ways in which she continually pushed against fixed boundaries, assuming an independence that she saw as a legacy from her ancestors" (qtd. in Carpentier). Like Glaspell, Claire's family is part of a tradition of people who left one place to venture to another and who called it a "new world," though it was only new to them. Just as they created "New England" as a blend of where they had originated and the world they had found, Claire is a unique blend of her family's history and her current context. New England enters the plant-Claire blend when Claire is called the "flower of New England" (Glaspell 32). The history of the "New World" then merges with information from the input space of plant life; what the reader already knows of New England and its colonization is now an input space connected to Claire and to flowers.

Like a plant that enters a new ecosystem, the pioneers adapted to a new environment, and the environment adapted to them. This adaptation, though it is a natural process, is also plant life's version of experimentation; in order to adapt to new surroundings, the plants must break out of their existing forms. The resulting "New World" was neither Europe nor the native land they had originally encountered; it was a unique blend of Europe, the America that existed before colonization, and an emergent structure that was neither one nor the other. By "running the blend" in this way, we can begin to make sense of the American experience in reference to Claire and her plants. The tension that arises within the blend communicates to the reader what many Americans were feeling at the time of the play's creation – an anxiety related to questions of belonging, identity, and experimentation. New ideals, technologies, cultures, and social structures were entering the ecosystem of American society and both the ecosystem and the new structures were in a state of experimentation, progressing toward new ways of being that had not yet been defined.

Glaspell's work has "an aesthetic fundamentally grounded in the dichotomy between migration and homestead" (Carpentier). Claire explains that "What has gone out should bring fragrance from what it has left. But no definite fragrance, no limiting enclosing thing. I call the fragrance I am trying to create Reminiscence" (Glaspell 32). While Claire longs for the plants to be new, she desires to give them "reminiscence." Thus, they are both rooted and uprooted throughout the narrative and exist in a state of perpetual uncertainty. Joseph R. Ugo argued that "the central theme, the overarching myth, the single experience that defines American culture at its core is migration" (qtd. in Carpentier). Glaspell's work reflected this "migratory consciousness" (Carpentier), as well as the concerns that came with migration. Claire's desire for "outness" (Glaspell 33) is continuously in conflict with Harry and Elizabeth's desire to remain in one place. Elizabeth tours Europe, but her experience only reinforces her love for her home (44), and Harry condemns Tom's traveling by proclaiming, "If you have a place – that's the place for you to be" (35). This tension between migration and homestead is established in the very first scene, which revolves around the displacement of the male figures; the men of the household are displaced when Claire turns all of the heat to the greenhouse where the men must have breakfast.

Claire cannot escape both the history of and desire for migration, yet she resents the men's migration to her domain because their migration has caused the temperature to drop, placing the plants in danger. The migratory consciousness in this play, then, revolves around an intricate and contentious blend. First, there is the input space of the plants. Plants that enter into a new ecosystem can be at risk themselves and they can pose a threat to the balance of the ecosystem, just as the men who are transplanted to the greenhouse pose a threat to the flowers that are housed there, and they feel threatened by the plants' existence because it is new and strange to them. When Harry complains about the heat being redirected to the greenhouse, he tells Anthony, "We can't all live out here" (Glaspell 41), meaning that the greenhouse is not suitable as a home for them. Anthony responds by stating, "Indeed you cannot. It is not good for the plants." Both parties have reason to be in the greenhouse, and both have concerns about the men's presence there, just as a new species in an ecosystem may be in danger for its own survival and other species could be endangered by it. Their displacement is equally problematic to the men and to the plant inhabitants of the greenhouse. This displacement and the tension it causes illustrate some of the tensions that arise in a society that experiences a rise in immigration, such as America at the turn of the century. Immigration had increased since 1820 and was expanding to include varying ethnic backgrounds, bringing with it political and social concerns about the unity and identity of the American population (Hill). In such a time, questions of resources and belonging arise.

The men's invasion of Claire's space creates a conflict grounded in questions of belonging and accommodation. The men have entered into an ecosystem as a foreign species and now they must adjust to the new environment, it must adjust to them, or they must leave. The ecosystem metaphor illuminates some of the conflicts between characters in the play, and between factions of American society. In any ecosystem, there is a balance between codependency and the survival of the fittest, an individualistic instinct. On one hand, living beings in an ecosystem rely upon one another's existence and, on the other hand, they compete for resources that will aid in their survival. Both codependence and self-interest, then, are part of the experience of living within an ecosystem. Claire has mutually beneficial relationships with the people and plants in her ecosystem, but also poses a threat to each of them. As she does when hybridizing her plants, she takes pieces of each of the people in her life; but just as she uproots and destroys the Edge Vine when it no longer serves her purposes (Glaspell 48), she destroys relationships when they no longer serve her needs. She is always situated physically near others, but she does not allow these relationships to define her. She is biologically a mother to Elizabeth, but emotionally a mother to her plants. She is legally the wife of Harry, but sexually involved with Dick and emotionally committed to Tom. At the height of the drama, Claire chokes Tom Edgeworthy when he attempts to contain their relationship within an existing form and therefore no longer fulfills his purpose in her life (72). Glaspell's decision to use strangulation as the method of killing further identifies

Claire with her plants; she, like plants, 'chokes out' the being closest to her when he attempts to take what she needs, when he becomes a threat to the survival of the identity and the otherness she desires. She is continuously on the verge, situated between community and isolation, between codependence and survival, and between her collective and individual identity.

Like America and the feminist movement, Claire is attempting to find a way to become a "higher type" of person, yet struggling to do so while maintaining human relations. Harry first complains about his wife's isolation by telling Anthony, "I too need a little attention" (Glaspell 26). Tom, too, begs her, "Don't go where we can't go" (71). Claire must live within the same realm as the men in the play, but attempts to find her creative self have caused her to neglect the duties required of a person living in a collectivist society. Since "Claire has rejected the roles of wife, mother, and mistress that are open to her and rebels against the suppression of self that society would enforce upon a woman, only to discover that the penalty is alienation" (Ozieblo 117), her isolation primarily results from her rejection of existing structures within which she could find solidarity. Claire is currently part of her environment, but is slowly working her way out of the ecosystem that requires her to live in codependency. Breaking out of the ecosystem, however, would mean being "lonely out in what hasn't been" (Glaspell 32), and Claire is trapped once again between homestead and migration, between codependency and isolation.

Questions of solidarity and isolation were also prevalent within the feminist movement of the time. Liza Maeve Nelligan claims that "The two plays that best reflect the dramatic developments in American feminist thought early in the century are *Trifles* and *The Verge*" (Nelligan 86). Women were on the verge of becoming a new form and both women and their surroundings would need to adapt in order to keep the ecosystem in balance. While *Trifles* presents the triumph of "a unified female morality and world," *The Verge* emphasizes individualism and rejects old notions of womanhood (86). The shift in Glaspell's representation of womanhood from the communal experience to the individual is one that reflects a central paradox of the feminist movement: "how to achieve a healthy sense of individualism without losing some of the more valuable qualities of 'womanhood,' particularly the sense of community that had informed *Trifles*" (92). At the end of the play, Claire has succeeded in forming a new plant and finding her way "out" through the murder of Tom. She is, however, completely isolated. She admits to the murder by stating, "I did. Lonely" (Glaspell 73) and rejects Anthony's offer to take her guilt so that she can continue her work. Claire, then, can only push the limits of her work and her existence so far. In this indeterminate ending, she has succeeded in her present experiment but cannot go further to give her Breath of Life the Reminiscence (32) she desires. The play ends with a woman who has broken the barrier, but is now isolated and cut off from any further successes. Just as she predicted Breath of Life would be, she is "lonely out in what hasn't been" (32).

By the end of the play, the two goals she had set, to be "out" and to create Breath of Life, are accomplished and they leave her unable to accomplish more and unsure of what more exists. As happened in the case of the suffrage movement, the primary goal had been reached and Claire is left with no certain sense of purpose. Once women gained the right to vote, they had to find ways to advance their movement further, but lacked a clear objective. Barber explains, "Her [Claire's] mutating plants provid[e] a clearly symbolic reference to the tenuous position of women in American society" (101). Glaspell uses Claire's character as a type of experimentation; she "is neither idealizing the superwoman nor is she parodying the fanatical feminist; instead, through Claire she is showing us what it feels like to be the Other, the alien, to be [Hélène] Cixious' newly born woman'" (Noe qtd. in Barber 104). It is in otherness that Claire connects her identity to that of Breath of Life. She claims, "Breath of Life is alive in its otherness" (Glaspell 30), finding life not only in the creation of new forms, but also in the identity of the Other as a possible definition for womanhood.

Claire has adopted both male and female roles; she controls the men's identities within the play (they are husband, lover, soulmate) and has the power to kill, but her femininity and sexuality are consistently present in the dialogue of the characters. Gainor claims that, "as a figure who has taken on male privilege and rejected the conventions of feminine behavior, Claire fits the model of the high modernist who scorns and transvalues traditional expectations" (158). The play, however, does not offer a singular, contained definition of womanhood. Like the "tower that is thought to be round but does not complete the circle" (Glaspell 49), Claire is a traditional representation of womanhood – on the surface, she appears to be a mother, wife, and mistress – but, at the same time, she does not fully "complete" these representations. She refuses to attend to her sister, who advises her that "A round tower should go on being round" (49) and that she should be the woman that she was "meant to be" (50). Rather, like the feminist movement, Claire and her tower evade boundaries and rely upon experimentation and imagination. The greenhouse, the primarily feminine domain, is introduced as "a laboratory" (26), a place where Claire tests new forms of plants, where Glaspell tests new forms of drama, and where women may begin to test new forms of selfhood. The play resonates with the feminist movement's desire to create women who are "something that has never been before: unconditionally free to choose one's life patterns" (Nelligan 91). Claire's choice to "smash" life and find her way "out" at the end of the play does not, however, display a woman who is so "unconditionally free." She has committed murder and now has two choices; she must submit to social structures of law and morality and be punished for her act, or she must rely upon Anthony, a male, to take her guilt. She is, therefore, still constrained by social structures and dependence upon a male figure.

This dependence makes the blended space of plant-Claire-Glaspell-woman particularly interesting. Most plants have both male and female reproductive organs and rely on self-pollination as their method of reproduction. In Claire's greenhouse, she is responsible for the creation (or birth) of the new plants; she pollinates and nurtures them, symbolically being both mother and father, just as a plant is both male and female. Like her plants, Claire has both male and female characteristics. She rejects the role of mother to Elizabeth but nurtures and "has given her heat to" (Glaspell 27) her plants, as though she were their mother. When Claire directs all of her heat to the greenhouse, Anthony reports that she woke up early because "Harm was near, and that woke her up" (27). Like a mother who can sense a threat to her children, Claire is woken by the threat to her plants. Just as her plants are meant to be something that has not been, she is still a woman, but is on the verge of becoming the New Woman, the woman that is outside of what women have been.

The ability to achieve otherness, however, requires not only creation, but destruction. *The Verge* "reflects the era's profound response to the chaos and destruction of this cataclysmic event [World War I]" (Gainor 147). Glaspell, like Claire, is exploring how the war may allow one to "Open the door to destruction in the hope of – a door on the far side of destruction" (Glaspell 41). Claire continuously expresses her desire to "smash something" and claims that she only married Harry because she "thought he would smash something" (39). She accuses Harry: "You think life can't break up, and go outside what it was? Because you've gone dead in the form in which you found yourself" (33), suggesting that it is only from disorder and destruction that something new can be born. She and her plants serve as the impetus for disorder in the lives of her family members, lovers, and friends. They must break the forms of their origins in order to embrace "outness – and otherness" (33). Glaspell pushes her drama so far beyond the boundaries of social order that her protagonist must revert to violence, an extreme form of disorder and unrest, in order to get "out." It is not the violence, however, but the experience of "where revolvers can't reach" (67) that takes one out. Claire claims that she is "too interested in destruction to cut it short by shooting." In *The Verge*, Glaspell is exploring the destruction that comes after the war – the neurosis, the breaking up of social constructs, the "otherness" (33). The violence of war had the potential to "smash something" (38), but it never quite fulfilled that potential because new forms and social structures were not born of the destruction. Claire attempts to achieve what the war could not when she murders Tom, but her motive, like her character, is ambiguous.

Just as Claire's character resists coherence and containment within a singular definition, the conclusion of *The Verge* denies any unified or definitive interpretation. After murdering Tom, Claire claims to be "out," and begins to recite the traditional hymn "Nearer My God to Thee." In one sense, she has reverted to the conventional depiction of suppressed woman; when the one man who could potentially love her in an unconventional way begins to refer to her as a possession, her only choice is annihilation. He has become "too much" and "not enough" (Glaspell 73). He is on the edge, on the verge, but does not find "outness" (33). While her actions have, throughout the play, been characterized as hysterical by the more conventional characters, "Claire talks us through her dawning understanding of her actions, much as any tragic hero, surrounded by the bodies of his beloved dead, talks us through his final self-revelation" (Malpede 123). She is fully aware of herself but still unable to create a language that expresses her newfound identity. Instead, she reverts to a classic, conventional hymn to express herself. She has become fully other; she has both created and destroyed life and now is "nearer, my God, to thee" after committing murder. She has broken conventional boundaries of life and death, sin and salvation. Glaspell may be suggesting that women could find a way to use convention, as Claire uses the hymn, yet not be subject to it. In Act I, however, Claire claims that "To perversion too there is a limit [...] If one ever does get out, I suppose it is – quite unexpectedly, and perhaps – a bit terribly" (Glaspell 31). In the end, she has found this to be true. She is "out" but it is "a bit terribly." The ending may be read as tragic in terms of the drama and its larger social implications. Claire has no choice but annihilation, a common ending for the modern woman who cannot "out" from existing structures. In this way, "*The Verge* haunts us with the possibility that alterity holds truths just out of our reach" (Gainor 169). Claire voices this notion with her claim that people, like plants, must "go mad – that life may not be prisoned" (Glaspell 39). Ozieblo believes that "in *The Verge* Glaspell's relentless investigation of the secrets of our civilization reveals tragedy to be the only possible outcome" (Ozieblo 117). These truths, though they may lead "out," "haunt" us with the possibility that the alternative, outness, is "a bit terribly" (31) realized and at the same time necessary if the human race is ever to create new forms and find "what has not been."

The plant blend, however, offers a unique perspective within this discourse of destruction and violence. Plants can create other plants like themselves through reproduction; however, it is only after they are destroyed that they have the potential to break from form and become something new. They must first die and decompose and then the elements of the plant can become nutrients for a different species of plant or animal life; the elements of the plant can then take a new form. Claire explains:

> Plants do it. The big leap – it's called. Explode their species – because something in them knows they've gone as far as they can go. Something in them knows they're shut in to just that. So – go mad – that life may not be prisoned. Break themselves up into crazy things – into lesser things, and from the pieces – may come one sliver of life with vitality to find the future. How beautiful. How brave (Glaspell 39).

The emergent structure of the human-plant blend enables Claire to talk about humans and plants as having the same potential to "break up," which comes from the plant input, but also to "go mad," which comes from the human input. In Claire's conceptual integration of plant-self-woman-American, this ability to take a new form may be the plants' "door on the far side of destruction" (Glaspell 41). This blend partially explains Claire's claim that she would "rather be the steam rising from the manure than be a thing called beautiful!" (72). Understanding Claire's claim this way adds to the conflicting characterization of Claire. While she is described as a self-interested and isolated woman throughout the play, here Claire may be admitting to a part of her character that has been hidden, admitting to a desire to give her life to nourish new life.

If Claire allows this metaphor to guide her decision making, her murder of Tom is not simply a fit of rage, but rather an attempt to help him "break out" when he is "going back home" (Glaspell 29) and becoming conventional. Like the Edge Vine and Elizabeth, Tom Edgeworthy has come close to otherness, but "doesn't want to be – what hasn't been" (29). Claire must give him the same opportunity that she has given to the Edge Vine – to "break up and go outside what it was" (33). The blend, then, gives new meaning to the ending of the play and the seemingly hysterical murder of Tom. It also brings a moral dilemma to the drama; the reader can no longer assume that Claire's actions were purely selfish or purely hysterical.

Using the plant metaphor to guide our understanding of Susan Glaspell's *The Verge* allows for a nuanced and fluid interpretation of the play that sensitively and accurately conveys the dynamic experience of being an American, and specifically a feminist artist in America, in the early twentieth century. It opens the text to a reading of the characters as belonging to an ecosystem and – in doing so – illustrates some of the conflicts and tensions created in the play as being the natural results of ecological changes. As new species are introduced, one of three changes must be made; either the new species is destroyed by other elements in the ecosystem, the ecosystem adapts to the species, or the species migrates again to a new and more suitable environment. Through the plant-human-New Woman-artist-American blend, Glaspell experiments with all three options for the New Woman and the New American the American society. Her experiment is inconclusive, but valuable in its indeterminacy. Through *The Verge*, Glaspell demonstrates the value of artistic experimentation in testing new forms of identity and social structures without risking the tragedy and annihilation that could result from a failed social experiment.

Through such artistic expression, she provides her readers with a way to exist in liminality, to imaginatively live on the verge of a new form, and ultimately to consider "truths just out of our reach" (Gainor 169).

Works Cited

Barber, Rytch. "American Expressionism and the New Woman: Glaspell, Treadwell, Bonner, and a Dramaturgy of Social Conscience." *Disclosing Intertextualities : The Stories, Plays, and Novels of Susan Glaspell.* Eds. Martha Celeste Carpentier and Barbara Ozieblo. New York, NY: Rodopi, 2006. 93-113. Print.

Carpentier, Martha C. "The Deracinated Self: Immigrants, Orphans, and the 'Migratory Consciousness' of Willa Cather and Susan Glaspell." *Studies in American Fiction* 35.2 (2007): 131-57. Web. 3 May 2010.

Fauconnier, Gilles & Turner, Mark. *The Way we Think: Conceptual Blending and the Mind's Hidden Complexities.* New York: Basic Books, 2002. Print

Gainor, J. Ellen. *"The Verge." Susan Glaspell in Context : American Theater, Culture, and Politics, 1915-48.* Ann Arbor, MI: University of Michigan Press, 2001. 143-169. Print.

Glaspell, Susan. *"The Verge." Four Plays. Trifles; the Outside; the Verge; Inheritors.* Teddington, Middlesex: Echo Library, 2007. 25-74. Print.

Hill, Howard C. "The Americanization Movement." *The American Journal of Sociology* 24.6 (1919): 609-42. Print.

Johnson, Mark & Lakoff, George. *Metaphors We Live By*. Chicago: The University of Chicago Press, 2003. Print.

Malpede, Karen. "Reflections on *the Verge.*" *Susan Glaspell: Essays on Her Theater and Fiction.* Ed. Linda Ben-Zvi. Ann Arbor, MI: University of Michigan Press, 1995. 123-127. Print.

Nelligan, Liza Maeve. "'The Haunting Beauty from the Life We've Left': A Contextual Reading of *Trifles* and *the Verge.*" *Susan Glaspell: Essays on Her Theater and Fiction.* Ed. Linda Ben-Zvi. Ann Arbor, MI: University of Michigan Press, 1995. 85-103. Print.

Ozieblo, Barbara. "Suppression and Society in Susan Glaspell's Theater." *Susan Glaspell: Essays on Her Theater and Fiction.* Ed. Linda Ben-Zvi. Ann Arbor, MI: University of Michigan Press, 1995. 105-117. Print.

"What is Americanism?" *The American Journal of Sociology* 20.4 (1915): 433-86. Print.

Surviving the City: Resistance and Plant life in Woolf's *Jacob's Room* and Barnes' *Nightwood*

Ria Banerjee

Abstract: In *Jacob's Room* (1922) and *Nightwood* (1936), Virginia Woolf and Djuna Barnes use plant life to express a profound ambivalence about the masculine-inflected ordering functions of art and morality. They show that these processes codify lived experience and distance it from the feminine and sexual. To counter this turn towards the urban inauthentic, both novels depict non-urban spaces to upend conventional notions of usefulness. They fixate on evanescent flowers, wild forests, and untillable fields as sites of resistance whose fragility and remoteness are strengths. In *Jacob's Room*, I argue that the eponymous protagonist is destroyed by his conventional education and morality, trapped in a room he cannot escape. In *Nightwood*, the central couple flees a similar ideological room, leaving for a remote farm where genuine feeling is momentarily possible.

Literary modernism is so much about urban life that it is surprising how much two novels set in European capitals of the early twentieth century invoke the mute and vegetable. Both Woolf and Barnes use plant life to express a profound ambivalence about the ordering functions of education, art, history-making, and morality which codify lived experience and distance it from the vital, feminine, and sexual, recording and representing inauthentic versions of reality. To counter this, both novels situate themselves in the marginal and upend conventional notions of usefulness. They fixate on evanescent flowers, wooded parklands, and untillable fields as sites of resistance against the censoring and ordering functions of art, and these natural elements serve as reminders of the need for more authentic representation of lived experience. In what follows, I take up the question of the Greeks and education in *Jacob's Room* (1922) to show how the eponymous protagonist is shaped and limited by the same; in the second section, I turn to *Nightwood* (1936) as the ultimately more hopeful text that depicts a central couple's flight from the city into a private, marginal space where it is possible to avoid definitions in favor of genuine feeling, if only for a moment.

The Modernism of Marginality: The Flower Girl and the Beacon of Ancient Greek

Foregrounding the confrontational nature of Barnes' novel, Jane Marcus writes that *Nightwood* engages with a "modernism of marginality" (88) that refuses the established conventions of the heteronormative Judeo-Christian world. It is only through absence that it figures "the authoritarian dominators of Europe in the thirties, the sexual and political fascists" (86) who dominate cultural and historical attention elsewhere. *Jacob's Room* is also intimately informed by a resistance that confines these "dominators" to its textual margins. Through attention to real and symbolic plant life, both novels construct a marginal modernist view that self-consciously circumlocates all contemporaneous political upheavals. Yet, the Great War is central to *Jacob's Room*, even more so perhaps than for *To the Lighthouse* (1927), since the loss of Jacob in combat is the core around which the novel's scheme of textual and formal innovations arises. Although the war is present only through absence, the text dwells upon other markers of prevailing authoritarianism[1]; in particular, it can fruitfully be read as a sustained

[1] Mentions of war brewing, although central to the tragic structure of the text, are buried under veiled references to smaller political events: for instance, although doomed from the opening pages by dint of his proleptic last name, Jacob Flanders is never shown to enlist in the Great War. Instead, the passage of time in the novel and his brewing desire for action are hinted obliquely as he reads The Globe one evening: "Jacob took the paper over to the fire. The Prime Minister proposed a measure for giving Home Rule to Ireland. Jacob knocked out his pipe. He was certainly thinking about Home Rule in Ireland – a very difficult matter. A very cold night" (98). Jacob's mind might be wandering at the time of reading, on Florinda's particular deception as I suggest later (that "certainly" adds a hint of sarcasm), but this piece of news pegs the date of Jacob's reading to early-1912, when the Home Rule Bill was introduced to Parliament. As the question of Irish Home Rule never enters the text again, and knowing it was completely set aside with the advent of the War, this oblique reference is primarily to situate the reader in time without the usual markers. Vara Neverow notes another way that Woolf includes the War: through mentions of domesticated horses and cavalry (2010). Building on Jane de Gay and Laura Doyle's comments, she further relates the emasculated, doomed horses of the army with "transgressive and secretive sexuality" (119), specifically aligned against the homogenised and censored world that Jacob tries to inhabit.

critique of upper class education as embodied by the Oxbridge tradition, whose propagation of Ancient Greek texts as an escape from the present and aggressively monkish "cloistered" (82) attitude come under fire through those excluded from it: the poorer, sillier, sexualized, and feminine. This essay looks particularly at Florinda as a locus of this critique, named for her flower-like purity but appearing tawdry in the clinical light Jacob has been taught to live in. Ephemeral in the text, plucked, displayed, and soon discarded, without her Jacob's life would nevertheless have been a "different affair altogether" (83).

Real truths are necessarily hidden and fleeting in *Jacob's Room*. Almost exactly midway through the novel, a novelty item popular in the early 1900s, little paper flowers which open upon touching water, make their way into the dining tables of the novel. At the end of the meal, floating in bowls of water or sinking down to the bottom, these little novelties provided opportunities for jokes and flirting among the company, or were taken as omens and signs in developing love affairs. They were thus directly responsible for "the union of hearts and foundation of homes" (83) in some cases. However, these paper flowers did not oust "the flowers of nature." Roses, lilies, and carnations continued to be popular because "real flowers can never be dispensed with. If they could, human life would be a different affair altogether. For flowers fade [...] [until they become] not fit to be seen" (83). In their decay, these "flowers of nature" become more than mere arrangements. They are rearticulated as memento mori which prophesy Jacob's death and emphasize the tragedy that this young man who "might become stout in time" (153), actually cruelly runs out of it.

Flowers are everywhere in Woolf's work, as a recent exhaustive critical intervention points out[2]. Bonnie Kime Scott writes that in Woolf, there is a move towards unified multiplicity that distances itself from the ordinary conception of nature and culture as being opposed to each other. Borrowing the term "naturecultures" (4) from Donna Haraway, she suggests that represented images of the natural "form a strong relation to the primordial" (5). Jacob experiences the interconnectivity of life, sex, and death on his first holiday to the beach as a child, but this primordial triad is dissolved through years of

---

[2] Elisa Kay Sparks, "Virginia Woolf's Literary and Quotidean Flowers: A Bar-Graphical Approach" from the Twentieth Annual International Conference on Virginia Woolf, 2010

systematized education, first Latin with Mr. Floyd, then Rugby, Cambridge, and finally the learning within the "enormous mind" (Woolf 108) of the British Museum. Woolf's recurring flowers, fields, and natural motifs are set up in opposition to "deconstruct patriarchal ideas of power and domination" as represented by these institutions, "and at least briefly defy spiritual defeat and death" (Scott 5).

Florinda is another floral manifestation in the text, Jacob's sometime lover and an artist's model who lives on the fringes of social and sexual respectability. Named so by a painter who "wished it to signify that the flower of her maidenhead was still unplucked" (77), Florinda herself is almost forgettable—she has no surname, and she drifts unmoored through London with the old and unreliable Mother Stuart as her only confidante. Florinda's thoughts are hardly as weighty and charged as those of the other women: Mrs. Durrant, Betty Flanders, and Mrs. Jarvis display depths of passion and insight beyond her. Fanny Elmer is more educated, taking up a Fielding novel on Jacob's recommendation (he gives Florinda a Shelley poem that she barely gets through (78-9)). Jinny Carslake is more confident and capable, a New Woman reveling in her urban freedoms, and Sandra Wentworth Williams is both older and more sophisticated.

Florinda is one extreme of unschooled womanhood whose opposite is Miss Julia Hedge, "the feminist" who watches Jacob copying out a passage of Marlowe in the British Museum (106-7) and experiences a frustrated irritation at him. Julia Hedge's reaction is not unlike that of the narrator in *A Room of One's Own* who also finds herself marginalized at the Museum because she is unable to come to any easy conclusions about her topic through research. Instead, she draws a doodle of the angry Professor von X (*ROO* 31) in her notebook, a sketch that is set up in opposition to the neat lists of notes compiled by the scholar working next to her. Julia Hedge is the locus of tutored feminine knowledge in this text, but in her own way she is as incomplete as Florinda who exists as simply as a cut carnation in someone's drawing room.

Susan Harris has written astutely about the politics of irony and parody in Woolf's novel, noting that it self-censors Jacob's sexual life to comment upon the functionings of power. For Harris, drawing on Foucault, the novel is complicit in Jacob's disgust with Florinda precisely because her low stature "makes it possible for sexuality (and anyone identified with it) to be apparently 'banished from reality'"

(423-4). That this artificial (censored, then normalized) reality is created and sustained by Jacob's educational background makes it possible to extend Harris' reading to suggest that Florinda (and the frustrated Julia) are at the opposite pole of value from Jacob's old professors, emblematic of particular social and economic class positions. The struggle of Florinda against Professor Sopwith, that she will inevitably lose, is a larger struggle between two principles of valuation. By highlighting the throwaway nature of Florinda's tragedy as compared to the aching loss of Jacob that suffuses the text, the novel sharply questions the standards that allow such a simple division of events into the macrocosmic (i.e., public, political, death at war) and the microcosmic (personal, domestic, an unwanted pregnancy). Hence, Florinda and the professor are implicated in Woolf's critique of the tyranny of plot conventions that writers are subject to, conventions that are themselves the result of a system of value.

In "Modern Fiction," Woolf argues emphatically for a new kind of "spiritual" literature (her example is Joyce's *Ulysses*, then appearing in episodic form) to counter the "materialist" urge in writers like Arnold Bennett and John Galsworthy. She suggests that "the proper stuff of fiction is a little other than custom would have us believe it" (*CR* 150), and fiction that accurately engages with the real must accommodate this. She is critical of all attempts to impose an external order on the chaotic stuff of lived experience through literary conventions of ordering, such as the narrative arc of the bildungsroman. *Jacob's Room* refuses to foreground the war as a more conventional novel might because it refuses to "take it for granted that life exists more fully in what is commonly thought big than in what is commonly thought small" (*CR* 150). Instead, the novel searches for artistic freedom to do away with conventions of plot and genre (what one critic calls "art's ordering function"[3]) to reflect reality in truer and fuller ways.

[3] This point is suggested by J.Scott Bryson in his reading of *To the Lighthouse*: "While Woolf does appear to be open to, at times even longing for, the possibility of art's ordering function, a keen ambivalence also exists within the novel, as she points out both the limitations and the destruction that result from an attempt to use art to order the world" (593). In *Jacob's Room*, a text written five years previous, that "keen ambivalence" is more pointedly employed as subversive irony, and art's "ordering function" is shown to be ultimately powerless and dangerous, unable to save Jacob from any of the pains he wants to avoid.

This problem of ordering can be reiterated as a problem with surfaces – that instead of simply registering sensory information, an internal, learnt system of story-making interferes to impose a narrative on events. Consider, for instance, Mrs. Pascoe, a tenant of the Durrants who is described by some tourists walking by: "Her face was [...] hard, wise, wholesome rather, signifying in a room full of sophisticated people the flesh and blood of life" (54). To the tourists, her life in a lonely corner of Cornwall smacks of the quaint and romantic ("Look – she has to draw her water from a well in the garden," they say to each other, 53). The tourists perceive her as the other half of a dialectic they inhabit, constructing a story about the "wholesome," honest villager to contrast with their urban sophistication. However, the very next line exposes their essentializing for what it is: "She would tell a lie, though, as soon as the truth." They have been fooled by the story of the picture of Mrs. Pascoe, failing entirely to grasp any real truth about her[4].

Flowers resist this story-telling by being impermeable to it, existing alive or wilted, fit or "unfit" to be seen. Unlike the tourists who view Mrs. Pascoe through filters of tyrannical plot, Florinda reaches an essential truth about Jacob almost as an afterthought: "Jacob. You're like one of those statues," she declares "dreamily" (80), comparing him to the Elgin Marbles in the British Museum. Francesca Kazan has detailed how Jacob turns into a "corporeal icon" (714) in the narrative which performs and parodies his ossification. Eventually, even Fanny Elmer echoes Florinda's insight, visiting the statues like one obsessed, trying to recapture some essence of Jacob from the depths of her loss. Jacob's education seeps through him, turning him into stone, and Fanny's emotions recognize him in the Marbles, themselves fragmented and transplanted relics of a lost age.

---

[4] This concern with pictures that lie informs much modernist thought, as if the advent of motion pictures made modernist literature doubly aware of the ordering function and dynamics of power in a still image. James Joyce famously tried and failed to run a movie theatre; D. H. Lawrence has written scathingly about "photographically-developed perfection" that turns the eye away from true vision and turns "[t]he picture of me, the me that is *seen*" into the essential "me" ("Art and Morality" 1925, 165). In *Nightwood*, Dr. Matthew O'Connor makes several references to women presented as beautifully arranged pictures, and the argument against regulated arrangement – and therefore for messiness and multiplicity – is a crucial part of that book.

Florinda – the only poor, pregnant, single woman in this text and thus a symbol of the unlettered, biologically-determined, class-oppressed feminine – is more than someone for readers to pity. Presented in "collages of modernist fragments – bits and pieces, or a rapid series of apprehensions," she (and Fanny Elmer to a lesser degree) are "characters in crisis or survivors of trauma – outsiders in search of a survivable system" (Scott 9). She is a crucial part of the book's arguments for different structures of knowledge and learning to "[protect] oneself from civilization" (146). Although Jacob is lost, a victim to his education and his times, perhaps this text holds out "some hope that, by touching back to the primordial, the semiotic, sensual, or material [...] a new and different cycle of human nature... may arise" (Scott 10).

In his limited and patronizing way, Jacob thinks of Florinda as a naïf who "could no more pretend a feeling than swallow whisky," a "little prostitute" with "inviolable fidelity" (94). Earlier, looking at her, he decides that "[b]eauty goes hand in hand with stupidity" and finds her suddenly very vulgar: "In spite of defending indecency, Jacob doubted whether he liked it in the raw" (82). Jacob is drawn to her, but his Cambridge education has trained him to deny the sexual instinct – watching her, "He had a violent reversion towards male society, cloistered rooms, and the works of the classics; and was ready to turn with wrath upon whoever it was who had fashioned life thus" (82). His education, although excellent, has done little to prepare him for the world, and further, has given him a sense of entitlement and competence that is dangerous. It is this complacency that Julia Hedge cannot stomach, his "regal and pompous" assumption that the "flesh and blood of the future depends entirely on six young men" (107) and that he is one of them. He fits perfectly in the world of the British Museum, next to the Elgin Marbles; Florinda brings his inflated ideals down to touch the messiness of lived experience, serving as both reminder and critique of the artificial world to him represented by Cambridge.

Cambridge occupies a unique position in this text which is so minutely concerned with depictions of spaces. If London (the streets, Jacob's and his friends' rooms) and Scarborough (Betty Flanders' establishment, Dodd's Hill) are two symbolic spaces in which this novel develops, then Cambridge is the third point of that triangle. Even more than classical learning, it represents a conglomeration of social and sexual signifiers that together generate meaning in much the same way that Oxbridge does in *A Room of One's Own*. It represents a particular sequence of learning and form of access; Jacob, too, finds that "the inevitable sequel" to Oxbridge is the British Museum (*ROO* 25). As pointed out above, Jacob is romantic about his education, admiring Fielding but actually responding to his curriculum with the haphazard intensity more associated with a Byronic hero. The most nuanced response to Cambridge doesn't come from Jacob but Chucky Stenhouse, the "unsuccessful provincial" (41) studying with the Old Professors Huxtable, Sopwith, and Cowan.

The narrative voice is loaded with irony when it describes Cambridge as a beacon of knowledge, asking: "Is it not simple, or pure, or wholly splendid, the lamp of learning, since if you see them there under its light (whether Rossetti's on the wall or Van Gogh reproduced, whether there are lilacs in the bowl or rusty pipes), *how priestly they look*!" (39-40, my emphasis). This text depicts the sublime in the small epiphanies of Mrs. Durrant or Mrs. Jarvis, and is an indictment of the "priestly," male-dominated yet sexless atmosphere of Cambridge that is so precious to Jacob. Cambridge presents itself as the only and best source of enlightenment – "'We are the sole purveyors of this cake'" (39) – and it convinces provincials like Chucky Stenhouse with "[t]alking, talking, talking" that "everything could be talked – the soul itself slipped through the lips in thin silver disks which dissolve in young men's minds like silver, like moonlight" (40). Sopwith is hypnotic to the young undergraduates, like Huxtable or Cowan who sings Virgil and Catallus "as if language were wine on his lips" (41).

Nightwood provides a pertinent comparison here. At the philosophical crux of that novel, Dr. Matthew O'Connor and Nora Flood engage in their most intense argument about the nature of love, its inherent egotism, and the damaging power of expectations to shape and reify the beloved. Matthew eventually bows down in the face of love while Nora finds reserves of strength to understand and bear up her own burden, to love Robin without the accumulated weight of socialized expectations. When Nora bemoans her love for Robin, which thus far has brought her only pain, Matthew admonishes her: "There is no truth [or single ideal of love], and you have set it between you; you have been unwise enough to make a formula; you have dressed the unknowable in the garments of the known" (Barnes 145). Making a formula to respond to the ineffable is precisely the problem of education Woolf is most concerned with, whether at Oxbridge or elsewhere. To define and make rigid, to turn an idea (or, "the soul") in all its multiplicity into a singular sliver of silver that can be easily slipped into the mouth is the chiefest of ills. Cambridge stands in for a denunciation of dogmatic education in all forms that aim to dress "the unknowable" of lived experience in "the garments of the known." In Barnes' conception, Nora can only understand and accept Robin's love when she moves beyond the conventional, the spoken, the egocentric. The couple finds each other in the wilderness of her country estate; Jacob, by contrast, is lost.

New structures of education are required to resist the old structures of power. Jacob's contemporaries, however, are powerless to withdraw from the system they are implicated in[5]. Chucky Stenhouse is drawn to Professor Sopwith like a moth to a flame despite the embarrassment of being addressed by his first name. There is a petty-minded menace in the professor's calling him "Chucky," a reminder of hierarchies of class and education in his refusal to address the younger man formally. Still, Stenhouse persists. In those evenings with Sopwith, the professor twined "stiff fibers of awkward speech – things young men blurted out – plaiting them round his own smooth garland, making the bright side show, the vivid greens, the sharp thorns, manliness" (41). For plaiting and arranging what they have said in half-understanding, for giving them a clear and singular vision of manhood that they can believe in, these young men worship all he stands for.

It is only when Stenhouse buys his newspaper the next day and catches the early train to university that "it all seemed to him childish,

---

[5] Christina Alt's careful study of lepidoptery in *Jacob's Room* points to the complex ways that Woolf uses natural plant and animal motifs to comment upon Victorian ideals of self-improvement through education and their relationship to the colonialist impulse to categorize and catalogue the Other in order to render it safe and controllable. On the one hand, Woolf's use of plant motifs is related to subverting dominant hegemonies of thought and action. At the same time, we see Jacob himself keeping a collection of moths and butterflies, sometimes staying out so late that it is past midnight by the time he returns home. However, Jacob's passion is related to the Victorian pastime of bug-hunting as an improving activity for children, focused on collecting and classifying what they find. For Alt, this is crucially related to the imperialist impulse to capture and codify, and thus Jacob's childish activity reproduces the structures of power that will ultimately lead him to death. "The *renunciation of capture* is central to Woolf's literary project" (Alt 133, my italics), a kind of collecting that is inherently different from, and opposed to, the colonial project.

absurd; the chocolate cake, the young men; Sopwith summing things up." Sopwith's method of adding up all the knowledge he deems necessary and leaving out what he doesn't echoes the clinical manner of the researcher at the British Museum who so frustrates the narrator in *A Room of One's Own*. That reader made "the neatest abstracts, headed often with an A or a B or a C while [the narrator's] own notebook rioted with the wildest scribble of contradictory jottings" (*ROO* 30). Although the narrator is distressed, bewildered, and humiliated by her lack of precise conclusions, it is exactly that riotous, living disorder that is reality itself. Like the contemporary writers belittled in "Modern Fiction," Sopwith is a writer constrained, "not by his own free will but by some powerful and unscrupulous tyrant who has him in thrall" (*CR* 149), returning inexorably to the same conclusions he was himself taught. There is a taste of old hat in Sopwith's "vivid greens" so that sometimes his silver disks of knowledge "tinkle hollow, and the inscriptions read a little too simple, and the old stamp look too pure, and the impress always the same – a Greek boy's head" (41). And yet, Stenhouse acknowledges the material advantages that Cambridge can give, which is perhaps a little other than the gift of knowledge: respecting still, he vows to "save every penny to send his son there" (41).

Any woman meeting Sopwith, "divining the priest" whose worldview leaves neither room nor agency for her, "would, involuntarily, despise" (41) him in a parallel reaction to that of Julia Hedge towards Jacob. But simplification has its attractions: Old Miss Umphelby, who sings Virgil as well as the other professors, attracts fewer student followers than Cowan. What she would most like to say in elucidation of the text, the lived details of "men's meeting with women which have never got into print" (42), cannot be said. Her attempts to bring the ancients off their lofty pedestal are more truthful but ultimately disappointing to youths like Jacob who are searching for clarity from the masters, craving a distance between themselves and that ancient civilization that would turn the words of the dead language into a magical antidote against the present. The tragedy of *Jacob's Room* is this realization that no antidote or elixirs exist and the ineluctable modality of the visible must be confronted.

Florinda, flower-like, is the agent and locus of this confrontation. Immediately after Jacob thinks of Florinda as a clichéd prostitute with a heart of gold, helplessly truthful about her love for him, he sees her "turning up Greek Street upon another man's arm" (94) in a moment of delicious narrative irony. Racked with jealousy, Jacob has a bad night and a bad week: socializing, riding in the country while "cursing his luck" (101), and having tea with his mother's friends, the Misses Perry and Rosseter. But Florinda haunts him and makes him judgmental and caddish until he finds himself in a room with the real prostitute Laurette. Even she allows him to sustain his careful avoidance of the sexual and vulgar, and only when he leaves he feels "that quake of the surface... which threatens to spill the whole bag of ordure, with difficulty held together, over the pavement." He has the profound sense that, "In short, something was wrong" (105).

The weak and silly Florinda possesses Jacob in ways that neither Betty Flanders nor Clara Durrant, Fanny Elmer or Jinny Carslake can. She is the spiritual realist core of the novel who exposes the "something wrong" that Jacob wants to avoid by enveloping himself in Greek. Jacob's end remains determined, but "that quake of the surface" continues to affect Jacob even when he reaches Greece and falls helplessly in love with the married Sandra Wentworth Williams. To Sandra, he eventually becomes one of a series of books on her shelf, reduced to an object for egotistic possession (161), but through his love for her and their shared love of Greek, he begins to glimpse a world of sense beyond the monkish world of talk he is familiar with. Miss Umphelby's unspoken insight manifests itself, although too late to deter the flow of the narrative towards its final, impossible confrontation with loss.

Just arrived from Paris, flush with the chatter of his friends there, Jacob enters Greece constantly editorializing: "'You ought to have been in Athens,' he would say to Bonamy when he got back," and "[make a] comparison between the ancients and moderns, with some pretty sharp hits at Mr. Asquith – something in the style of Gibbon" (136). Already an enthusiastic essay writer, Jacob collects sights and impressions that he twines into garlands for himself, imitating not only the tastes but also the methods of his Cambridge professors. The landscape speaks to him but he is busy recording only his own voice. Only gradually he finds "how tremendously pleasant it is to be alone; out of England; on one's own; cut off from the whole thing" (141). He climbs to the top of lonely mountains in the intense Mediterranean afternoons, but instead of planning a letter, thinking about what to say to friends, or plotting the next impressive but dull essay, Jacob descends into a deep silence so that "[s]tretched on the top of the mountain, quite alone, Jacob enjoyed himself immensely. Probably he had never been so happy in the whole of his life" (144). In keeping with the ironic spirit of the novel, the Grecian afternoon landscape takes him farthest from his ideals and is perhaps closest to the dumb muteness embodied by Florinda.

The essay writer, the copier of passages from Marlowe, finds himself now robbed of words. Confronted by the "extreme definiteness" (148) of the Parthenon and those obdurate stones on which "the emotion of the living breaks fresh [...] year after year" (161), Jacob's egocentric security is rocked for the first time. He is in the grips of something bigger than a fleeting affair with a married woman. To himself, he calls it "this sort of thing" without being able to elucidate further. It's hard to write to his mother, and to Fanny he only sends postcards. Even with Bonamy, something stops him from asking his friend to rush to Athens and share "that uneasy, painful feeling, something like selfishness – one wishes almost that the thing would stop – it is getting more and more beyond what is possible." Cambridge has shown Jacob – like Stenhouse – that it is possible to climb ladders both of learning and social standing. It gives them the power of Greek against their deadening day jobs; only now, Jacob feels trapped by his circumstances in face of the Parthenon. Add to this tumult are his love for Sandra and the attendant problem of sex.

Jacob continues thinking in disjointed sensual fragments, severely discomfitured: " – the sight of Hymettus, Pentelicus, Lycabettus on one side, and the sea on the other, as one stands in the Parthenon at sunset, the sky pink feathered, the plain all colors, the marble tawny in one's eyes, is thus oppressive" (149). Betty Flanders has had similar trance-like moments upon Dodd's Hill, as does Mrs. Durrant in Cornwall with the boy Curnow at her side, and Mrs. Jarvis on her rambles on the moors at night. Even Sandra Wentworth Williams asks herself a profound "What for? What for?" (161). In dealing with "this sort of thing" that refuses to be quelled with simple common sense or reading the newspaper, in his own way Jacob is at his most feminized. He loses the accumulated baggage of his education and is confronted by "an unseizable force" that is lived experience. Like novelists, he can "never catch it" because "it goes hurtling through [his] nets and leaves [him] torn to ribbons" (156). Florinda is the true priestess of this unseizable force in her animal sexuality and untutored thoughts, her mawkish letter-writing and her dullness. Years after Jacob is gone, mad Jinny Carslake shows strangers her box of ordinary pebbles picked off the road, beribboned by that unseizable force herself. But looking steadily at the stones, she knows what Jacob only dimly senses, that "multiplicity becomes unity, which is somehow the secret of life" (131).

I began this section with the idea of *Jacob's Room* as well as *Nightwood* being implicated in the poetics and politics of a modernism of marginality. Multiplicity becomes unity in Woolf's text, as Jinny's pebbles remain discrete in marked difference from Sopwith's plaited garland, where all ideas are bent and woven into a dominant master pattern. It is significant that none of the feminine voices in the text meld together but remain singular even when they are related. Florinda and Fanny move in the same circles but, despite loving the same men (Jacob and the painter, Nick Bramham), do not see each other favorably. Clara almost betrays her mother to Mr. Bowley and only narrowly recovers herself (165-6). Betty Flanders "knew precisely how Mrs. Jarvis felt," even though "she never listened to her discontent" and usually counters the latter's deeply-felt restlessness with some homily or home remedy (91). Even for this closest of pairs, there is much left unsaid; for instance, when the two women walk the nighttime heath in a culminative moment, each remains locked in her private subjectivity. Betty Flanders wonders about the garnet brooch Jacob gave her which she lost; without intruding on her friend's hidden quest, Mrs. Jarvis watches her stoop to pick up something and thinks, "Sometimes people do find things" (132). In its allusive style and the way it insists on allowing each of these half-submerged lives its own subjectivity, *Jacob's Room* tries to find a moment of truth in the marginal as well as the marginalized.

Leaving for American Wilds: Silence and the Rejection of the Old World

While Jacob is en route to Greece and experiencing a concurrent rejection of language, the narrative voice muses on the relationship between human life and the environment to suggest, "Though the opinion is unpopular it seems likely enough that bare places, fields too thick with stones to be ploughed, tossing sea-meadows half-way between England and America, suit us better than cities" (144). Implicit in this statement is a move away from spaces chosen for their usefulness (defined commercially), as it articulates a movement towards marginal, feminized spaces and away from the masculine city where so many are ruled by the markings of the workday clock face. The final move from urban commercial centers to useless wilds that Jacob is unable to sustain is the driving impulse of *Nightwood*.

Nightwood is similarly implicated in the ways artificial narration imposes an order from without that doesn't fit lived experience. In Barnes' book as in Woolf's, there is a pervasive sense that we are "doomed all [our] days to write letters, send voices, which fall upon the tea-table, fade upon the passage, making appointments, while life dwindles" (*JR* 93). Both texts fixate on talking – *Jacob's Room* in the figures of Sopwith and Jacob, and *Nightwood* in the almost-unstoppable flow of Dr. Matthew O'Connor – and yet both have a deep-rooted consciousness that most "words have been used too often; touched and turned, and left exposed to the dust of the street," while those words that we really seek nestle "sweet beneath the leaf" hidden "close to the tree" and are only visible at dawn (*JR* 93). For Woolf, those sweet, new, hidden words hold out the possibility of true communication; Barnes goes a step further, showing those words to be tacit, embodied, unspoken. The closest that narrative can venture into describing that level of communication is to report the mimicked barks of a dog; it stands mute at the threshold with words it cannot reach as if confronted with the depths of the unconscious.

If Woolf's text can be read primarily as a struggle against the effects of an entrenched education, as I suggest, then Barnes' text is an indictment of notions of stable history imposed "upon time and open air" which also uses plant life to locate its resistance. *Nightwood* begins by likening its protagonist to an enormous, exotic house plant, and repeatedly conjures what Teresa de Lauretis describes as the *selva oscura* for its literal and psychological landscape. As she argues, the associations with Dante's more famous forest aligns Barnes' use of the trope of vegetation with another quest to leave behind disorientation in search of knowledge and truth. More concurrently, Freud's *The Interpretation of Dreams* "has reinscribed the trope of the journey in an altogether different dark wood" so that "the journey is henceforth interminable, reversible, discontinuous and intertextual" (de Lauretis 118). The journey in *Nightwood* has a different tenor from those previous masculine journeys; here the dark wood is not terrifying and uncanny, but the living space of freedom.

The positivism of this non-linear sequence (the lovers find each other, albeit perhaps momentarily, at the end) allows *Nightwood* to be read as strongly anti-fascist, reacting against the dominant master narratives shaking Europe in the late 1930s. Barnes' characters return to the dark forests of their minds and in their frankest moments are drawn to the woods closest to them in Paris, the Bois de Boulogne. In this section, I will focus on three sequences: Felix's provenance, his first look at Robin, and the concluding chapter (including Dr. Matthew O'Connor's breakdown), to suggest that *Nightwood*'s struggles against historical determinism locate useless plants and wooded spaces as sites of resistance and polysemy. Its warning against seeing in pictures recalls Jacob's petrification, and its final scene is a discarding of the ill-fitting garments of tutored language for a truer communion between the lovers.

At a critical juncture in the novel, midway through Matthew's second conversation with Nora, the former offers an insight that is central to this text. Selecting a story as if at random, he tells her of a priest, Father Lucas, who advised him to "be simple like the beasts and yet think and harm nobody" (140). This statement carries at its heart a central paradox: to be beastly and yet sentient, rejecting neither the animal nor human. It is a statement that recalls the simplicity of mad Jinny Carslake's box of pebbles, and carries the impulse to find unity within multiplicity. Robin is on this very quest to straddle the fine line between the human and animal, and yet while she remains in Paris, the closest she reaches to her essential nature, where the two naturally commingle, is through her contact with plant life. Before turning to Robin-as-plant, however, let us examine Felix and the problem of historical determinism against which Robin and Nora react.

Felix is born orphaned of a Jewish Italian father passing as a Protestant German nobleman, "heavy with impermissible blood" (5), and a Christian mother who unconsciously "stalked" (7) her husband's assurances about his family history even while she claimed to believe them. Guido's search for authenticity had been warped and twisted so that, instead of being "simple like the beasts," he embarks upon an unending journey for outward legitimacy by embracing his fictitious family history and attempting to continue a line he was always already dispossessed of. Jacob pursues and embraces an education that allows him to claim a place in the world rather better than his roots would otherwise allow; Guido senior and his son Felix fabricate their entire past in another manifestation of the same impulse. Ultimately, both attempts to find and fix one's place result in failure.

When Felix meets Matthew, the latter immediately diagnoses him as having "something missing and [being] whole" at the same time, "damned from the waist up" (29). He likens Felix to the poor legless Mademoiselle Basquette, inviting us (because his audience, the drunken Frau Mann, is barely listening) to draw the implication that Felix's search for legitimacy has halved him and left him helpless to the whims of powers beyond his control. Because he doesn't believe himself whole, he is liable at any moment to be violated by sailors of circumstance like the disabled Mlle Basquette. If Jacob is figuratively petrified by dogmatic education, Felix is halved and left incomplete by his adherence to a too-narrow conception of history. Recall the terms of Woolf's denunciation of "modern fiction" – Felix is the ultimate manifestation of an author in the thrall of "some powerful and unscrupulous tyrant" (*CR* 149) that insists that life exists more fully in the "commonly thought big" than in the small. His insistence on naming himself a baron and addressing his wife as "the Baronin" (retaining and normalizing the German term) is an act of ultimate self-renunciation, rendering both himself and Robin hollow. It is Felix's misfortune that he sees this giving up as its opposite, an act of supreme self-fulfillment; his punishment is that instead of finding a wife like Hedvig, who would be content to outwardly discipline her discontent, he marries Robin, a woman deeply engaged in escaping this same determinism. Robin recognizes the Mlle Basquette in herself and her personal quest to escape unitary towards the wholeness of multiplicity must come at the expense of her husband.

Barnes' description of Guido and Hedvig's house recalls Woolf's obsessive return to the Elgin Marbles: "The long rococo halls...were peopled with Roman fragments, white and disassociated; a runner's leg, the chilly half-turned head of a matron stricken at the bosom, the blind bold sockets of the eyes given a pupil by every shifting shadow so that what they looked upon was an act of the sun" (8). Later, the selfish passion between Robin and Jenny Petherbridge similarly locks them in place "like [sculpted] Greek runners, with lifted feet but without the relief of the final command that would bring the foot down – eternally angry, eternally separated, in a cataleptic frozen gesture of abandon" (76). In both cases, the narrative emphasizes that these figures are in a state of agonized waiting, expecting or relying on an outside impetus ("the sun") to give them direction ("what they looked at"). Felix tries to insert himself into an extant history thinking that doing so might lend him authenticity and authority. In a novel that so aggressively marginalizes authoritarian dominators, history is rendered as yet another (dominant) fiction, and Felix as a character who tries to enter a story that is not his. He is thus confined to being a mere "foot soldier of history" (Marcus 95), a ridiculous figure who cannot help bowing to whoever looks most distinguished in the cafes he frequents "with the abandon of what a mad man knows to be his one hope of escape, disproof of his own madness" (131). Despite achieving some insight into his marriage and the state of his son by the end of the narrative, Felix remains trapped by his old habits. To give them up at that advanced stage would be a too-great abnegation for a personality built on such shifting sands of fiction.

The love triangle created by Jenny Petherbridge lies outside the purview of this essay, but it is worth noticing how closely her dilemma of self-representation echoes Felix's. Where the latter chooses public history to insert himself into, she vampirizes and inscribes herself into the "commonly thought small" stories of other loves. Jenny hopes to replicate and surpass the relationship between Robin and Nora, and therefore also fails. To her, theirs is a story she wants to make hers, as if their relationship is a tangible thing to wrap around herself. She plays a role even when in love: "One inevitably thought of her in the act of love emitting florid *commedia dell'arte* ejaculations" (74), substituting genre clichés for what Woolf names "spiritual" realism. Like Felix, Jenny is uninterested or perhaps unable to create her own story and remains on the outside of her own life. Jenny is no Florinda; by immersing herself in other people's lives and stories, she drifts away from truth helplessly.

"Your devotion to the past," Matthew says to Felix as they are driving to the *Bois* one evening long after the separation of Felix and Robin, "is perhaps like a child's drawing" (119). And indeed, a child's simple sketch of the past, with its definite outlines and skewed perspective, is what Felix (also, Jenny) has been chasing. Although troubled, Felix agrees with Matthew: "My family is preserved because I have it only from the memory of one single woman, my aunt; therefore it is single, clear and unalterable... through this I have a sense of immortality" (119-120). But although the Judeo-Christian world is about separation, "*Nightwood* is about merging, dissolution, and, above all, hybridization" (Marcus 88). Felix's conception of domesticity is not only inappropriate for the needs of his estranged wife; within this novel, it denotes a dangerous rigidity.

In a parallel to Jacob's love for the Greeks, which is formed without consideration for their social and political history (*JR* 76) and culled from what he has haphazardly absorbed, Felix wants the unthinking absorption of routine, a childlike harking for stability established through familiarity. By the end of their conversation (significantly, held at the upscale restaurant in the *Bois* of Paris), Felix is momentarily unglued from his stasis. He admits to Matthew, "I wanted […] to find, if I could, the secret of time" but "that is an impossible ambition for the sane mind. One has, I am now certain, to be a little mad to see into the past or the future […] to know life." Accepting the loss of his wife and his son's ruined mind, he speaks for the first time with understanding about Robin: "[To know life] may also be the errand on which the Baronin is going" (130).

Robin herself is in America at this point, moving away from Jenny and towards Nora, who has left Europe to return to her ancestral estate, a bare place also "too thick with stones to be ploughed" (Woolf 144). It is fitting that the lovers, who meet in New York and travel east to Europe before this final journey doubling back westwards, return finally to a place resisting commercial usefulness. Turning their back on commerce, the triumph of the final scene is latent in the vegetable way that Robin is described at first. This first look of Robin has received much critical attention. Robin's appearance as a dancer arrested mid-step (38) can be read as a parody of Charcot's methods of hypnosis (Marcus), a rewriting of the birth of Venus (Capelli), and a parallel to the opening scene with Hedvig (Rupprecht), to name a few.

For the purposes of this paper, what strikes me most powerfully is the degree to which Robin has taken on the aspect of the "confusion of potted plants, exotic palms and cut flowers" (37) as she sleeps. Her body exudes a smell "of the quality of that earth-flesh, fungi, which smells of captured dampness and yet is so dry." Even her skin is "the texture of plant life," the frame beneath "broad, porous and sleep-worn," with the "troubling structure of the born somnambule" (38). It is almost as if she is conscious in her sleep—although they were having trouble rousing her, once Matthew and Felix arrive, she slips easily between speech and oblivion like one recuperating by being lost in dream. Marcus points out the historical connection between Bellini's nineteenth century opera and the concurrent belief that sleepwalking proved the existence of a spirit by showing that consciousness is not all (108-9). Robin's urge towards a personal version of Catholicism, her longing "to know life," are manifestations of an aggressively twentieth century reiteration of this eighteenth century belief in the existence of the soul. Her likeness to plant life distances her from the marbled, fragmented Felix; her organic soul, unlike his, is able to retreat and find a momentary peace.

But this first look at Robin is problematic too, because she is not simply of the texture of plant life. The two men, seeing her mad dancer's pose on the bed, immediately place her in a more familiar frame: "a painting by the *douanier* Rousseau" (38). There is a hint of ridicule in the French descriptive and mad laughter in the subsequent changed view of Robin as "thrown in among carnivorous flowers as their ration" in a place owned by "an unseen *dompteur*, half lord, half promoter, over which one expects to hear the strains of an orchestra of wood-winds render a serenade which will popularize the wilderness" (38). In contrast to the previous attention to surfaces[6] – her trousers, shoes, smell, skin – these attendant observations present a single picture that turns Robin into an actress in a melodrama of exoticism and music.

Hiding behind the palms out of decency, Felix is taken with this chain of signification (recall Woolf's parody of the tourists' view of Mrs. Pascoe). He is already a lover of the circus, mistaking its carnivalesque upending for the "particular *Comédie humaine*" (12) he was in search of. The circus performers "took titles merely to dazzle boys about town, to make their public life mysterious and perplexing" while "Felix clung to his title to dazzle his own estrangement" (14). Felix misses the parodic and campy elements in circus life entirely, instead, he goes going among them for "that sense of peace that formerly he had experienced only in museums" (14). As Jane Marcus discusses with attention to Nikka and Frau Mann, the camp aspects of

---

[6] Rupprecht notes this attention to surfaces to do a thorough anti-Nazi, resistant reading of *Nightwood*'s aesthetic: "Barnes insists on the importance of art and culture by depicting things in terms of their surfaces (rather than their 'essence'). The narrator frequently describes people submerged by the texture of their environment, as in the case of the trapeze artist Frau Mann […] Hedvig is [similarly] described in terms of her reproductive capacity, a capacity which seems, however, vital to the environment in which she appears" (102).

their performance is the locus of their power: shifting, evanescent, embedded in lies, and as hard to pin down as reality itself. But, like Woolf's tourists, Felix's eye converts their emphasis on surfaces into one of essences, seeing them as relics to be placed in yet another metaphorical museum. Similarly, when he sees Robin as a primitive exotic half-beast, half-woman controlled by an unseen circus tamer, "he felt he was looking upon a figurehead in a museum" (41)[7]. Grasping this picturesque impression, he pursues her in an inevitable slide into disillusionment that ends with Robin rejecting both her marriage and her baby with a feral, "I didn't want him!" (53). Her final "him" expands and envelops all – the infant Guido and expectations of maternity, Felix and his imposition of domesticity, and all manner of dominant aggressors to which she is subjected.

The central problem here, as in *Jacob's Room*, is that of clarity. It is perhaps not so bad that Sopwith is constantly "talking, talking, talking," but it is dangerous that he condenses and eliminates the different in favor of the same. Similarly, by missing the camp performative elements of the circus and Robin's essential plant-like nature, Felix turns both into homogenous elements of his own search for authenticity. As the narrative cautions, "The woman who presents herself to the spectator as a 'picture' forever arranged is, for the contemplative mind, the chiefest danger" (41). The clarity each achieves through this simplification is false (see also: footnote ii), and in Felix's case, by the imposition of a birthing on Robin, violent. Hence, when she gives birth "[a]mid loud and frantic cries of affirmation and despair," not the

---

[7] An interesting side note: Djuna Barnes is said to have approved of Joseph Frank's assertion about this novel's reliance on spatial form and apparent lack of sequential organisation. Instead of using time as an organising principle (which even *Jacob's Room* does, as the novel is largely sequential), Barnes uses space, weaving her text together through a "continual reference and cross-reference of images and symbols" (Frank 32, also Rupprecht 95). Here, the recurring instances of ancient, broken, decontextualized marble sculptures acquire a palimpsestic quality that allows us to see the couples Felix-Robin, Nora-Robin (first time), and Jenny-Robin as essentially manifestations of the same relationship based on ego and power, desiring to rid the beloved of agency, objectifying her into paralysis. Carolyn Allen asserts that by showing not only heteronormative, but also same-sex desire, the text creates this palimpsest of differences that produce "a doubled subjectivity of resemblance" (22-3). It is only the second iteration of Nora-Robin that breaks out of rigid causality and a new kind of love relationship emerges at the moment of the novel's closure.

child but "*Robin* was delivered" (52, my emphasis). This ultimate reduction to her biological function at the cost of all else stirs her out of the "stubborn, cataleptic calm" (49) that she assumed was her only power. It pushes her to reject motherhood as her first major assertion. Her wild search for Nora is the second, even more radical than this first.

Felix should have read his wife for what she said and left unsaid. Spying her one day in a small tapestry shop facing the Seine, he thinks he has discovered the secret behind her clothes, which were "of a period that he could not quite place" (46). This episode is quintessential Barnes, combining philosophical depth with a prosaic explanation of Robin's financial situation. To Felix, the mystery of Robin's clothes is solved as soon as he sees her "reflected in a door mirror of a back room, dressed in a heavy brocaded gown which time had stained in places, in others split, yet which was so voluminous that there were yards enough to refashion" (46). He places her financially and socially, then asks her to marry him and is surprised she accepts.

And yet, Robin's dressing style is an assertion against allowing herself to simply be converted into a series of significations. Recalling Walter Benjamin on translation, the folds of her dress cloak her true self like language does meaning; de Lauretis invokes Jean Laplanche to assert that "subjectivity may be understood as a process of self-translation, detranslation and retranslation" (120). Even when she is with Nora in Paris, she walks "in *formless* meditation, her hands thrust into the sleeves of her coat" (65, my emphasis) so that both her figure and her thoughts escape definition. In other words, "Meaning is oblique and camp, but it is not a [singular] secret to disclose; language's lack of transparency is, paradoxically, there for all to see" (Caselli 161). *Nightwood* parallels *Ulysses* in its use of "mixed metaphors, mixed genres, mixed levels of discourse from the lofty to the low, mixed 'languages' from medical practice and circus argot, church dogma and homosexual slang" (Marcus 88). In it, meaning escapes definition and remains immanent, in direct contrast to Felix's doomed method of self-fashioning.

This is why, ultimately, Dr. Matthew-Mighty-grain-of-salt-Dante-O'Connor (Barnes 87) suffers a violent breakdown despite his acuteness. He remains trapped in the world of "talking, talking, talking" even as he is instrumental in setting his friend Nora free. He reinscribes the Freudian world of sexuality, and pines for the reproductive function that *Nightwood* has already discarded (Rupprecht 107). "Why doesn't anyone know when everything is over, except me?" he cries in a drunken anguish, and the sad irony is that his "Mighty-grain-of-salt" is not ultimately large enough to allow him the freedom of Robin and Nora. He circles around the *selva oscura* trapped in a deterministic universe. Addressing an audience that is in turns mocking and pitying, he lays out the depths of his agony: "Now that you have all heard what you wanted to hear, can't you let me lose now, let me go? [...] it's all over, everything's over, and nobody knows it but me" (175). With the hindsight of history, this section is chillingly prescient as the old certainties emphatically vanish. The doctor clings to the past in his own transvestitism, dreaming of a happy domesticity where he "boils some good man's potatoes and tosses up a child for him every nine months by the calendar" (91).

But *Nightwood* calls for new ways, and Robin's return to Nora is a reminder that "the human condition is a sister- and brotherhood of difference, and that ideologies that seek to erase those differences and define only themselves as human are indescribably dangerous" (Marcus 118). It answers Molly Bloom's yes with "Robin Vote's no to marriage, no to motherhood, no to monogamous lesbianism. Robin's no is a preverbal, prepatriarchal primitive bark" (Marcus 96). It is ideologically opposed to fascist aesthetics even when the same are coolly extirpated from the text, its "structure of reflexivity, which is related to the figure of irony, [not allowing] any kind of totalizing vision, since by definition [totalizing] creates a split within the subject" (Rupprecht 96-7). Sexuality, which remains problematic at the end of *Jacob's Room*, is more happily resolved here, and the false dialectic between monkish learning and feminine knowledge dissolves in *Nightwood*'s highly-cinematic closing scene.

In *Jacob's Room*, history's debris are left in the European capitals to continue as they must, confronting the everyday with confusion and pain. By contrast, *Nightwood* offers the possibility of an organic holism at a distance from historically-sedimented urbanity. Matthew's vision of "now *nothing, but wrath and weeping*" (175) is countered by the two "Possessed" (the name of the brief final chapter) with an "excess of affect" (de Lauretis 137) that is so far beyond the bounds of signification that it can only exist, even if for a moment, in the desolate, disused chapel that Nora possesses. Plant life, which sustained Robin at the beginning of the novel, surrounds the couple in that final scene, offering hope that it is possible to survive the city. Unhappy Felix retreats from his insights but Nora and Robin retreat into them, becoming finally as wordless as vegetable life itself.

Works Cited

Allen, Carolyn. *Following Djuna: Women Lovers and the Erotics of Loss.*Bloomington, IN: IUP, 1996. Print.

Alt, Christina. "Virginia Woolf and Changing Conceptions of Nature" in *Virginia Woolf's Bloomsbury*. Eds. Gina Potts and Lisa Shahriari. New York: Palgrave Macmillan, 2010. Print.

Barnes, Djuna. *Nightwood*. New York: New Directions Press, 2006. Print.

Bryson, J. Scott. "Modernism and Ecocriticism" in *Modernism Volume 1*. Eds. Astradur Eysteinsson and Vivian Liska. Amsterdam: John Benjamin, 2007. Print.

Caselli, Daniela. "The 'Indecent' Eternal: Eroticism in Djuna Barnes' *Nightwood*" in *Modernist Eroticism*. Eds. Shane Weller and Anna Katherina Shaffner. New York: Palgrave Macmillan, 2012. Print.

De Lauretis, Teresa. *Freud's Drive: Psychoanalysis, Literature and Film*. New York: Palgrave Macmillan, 2010. Print.

Harris, Susan C. "The Ethics of Indecency: Censorship, Sexuality, and the Voice of the Academy in the Narration of *Jacob's Room*." *Twentieth Century Literature* 43.4 (1997). Print.

Kazan, Francesca. "Description and the Pictorial in *Jacob's Room*." *ELH* 55.3 (1988). Print.

Lawrence, D. H. "Art and Morality" in *Study of Thomas Hardy and Other Essays* ed. Bruce Steele. Cambridge: CUP, 1985

Marcus, Jane. *Hearts of Darkness: White Women Write Race*. Rutgers: RUP, 2004. Print.

Neverow, Vara. "The Woolf, the Horse and the Fox: Recurrent Motifs in *Jacob's Room* and *Orlando*" in *Virginia Woolf and the Natural World: Selected Papers from the Twentieth Annual International Conference on Virginia Woolf.* Eds. Kristin Czarnecki and Carrie Rohman. Clemson, SC: Clemson University Digital Press, 2011. Web. 2 July 2013.

Rupprecht Caroline. *Subject to Delusions: Narcissism, Modernism, Gender*. Evanston, IL: Northwestern Univ. Press. Print.

Scott, Bonnie Kime. "Ecofeminism, Holism, and the Search for Natural Order in Woolf" in *Virginia Woolf and the Natural World: Selected Papers from the Twentieth Annual International Conference on Virginia Woolf.* Eds. Kristin Czarnecki and Carrie Rohman. Clemson, SC: Clemson University Digital Press, 2011. Web. 2 July 2013.

Sparks, Elisa Kay. "Virginia Woolf's Literary and Quotidean Flowers: A Bar-Graphical Approach" in *Virginia Woolf and the Natural World: Selected Papers from the Twentieth Annual International Conference on Virginia Woolf.* Eds. Kristin Czarnecki and Carrie Rohman. Clemson, SC: Clemson University Digital Press, 2011. Web. 2 July 2013.

Woolf, Virginia. *The Common Reade: First Series.* Ed. Andrew McNeillie. New York: Harcourt, Inc., 1925. Print.

—. *Jacob's Room.* New York: Harcourt, 1923. Print.

The Smell of Cottonwood Leaves: Plants and Tayo's Healing in Silko's *Ceremony*

Ubaraj Katawal

Abstract: In Leslie Marmon Silko's novel *Ceremony*, a struggling World War II veteran, Tayo, who has lost his cousin in the war, finds out that his home has become unwelcoming. Facing a conjuncture of not knowing who he is and where he belongs, Tayo finds help in nature, especially in plants. He finds connection to the Earth and to his native culture by understanding how he is connected inextricably to the plants and animals around him.

In a world of crickets and wind and cottonwood trees he was almost alive again; he was visible.

Leslie Marmon Silko, *Ceremony*

All of Silko's work is infused with reverence for the natural world. Her "tellings" never lose sight of the fact that the earth was here first, along with the sun and the moon and other permanent powers.

Larry McMurtry, "Introduction" to *Ceremony*

Leslie Marmon Silko's *Ceremony* tells the story of a World War II veteran, Tayo, who suffers from "war fatigue," or posttraumatic stress disorder after he loses his cousin, Rocky, in the Pacific theater. Adding to Tayo's remorse for not having been able to bring his cousin safely back from the war, as he had promised to Auntie (*Ceremony* 67), is the death of Josiah, his maternal uncle, back at home in New Mexico while Tayo and Rocky were away fighting the war. Josiah had helped Tayo grow after his mother, Laura, abandoned him with the family at the age of four. When Tayo returns home from the war, he cannot help blaming himself for the deaths of the two most important persons in his life, besides old Grandma. He is sick, as a result. With the help of old Betonie, a Navajo medicine man, Tayo begins to slowly recover from the illness, but not before he takes several important steps, including a successful search for the missing Mexican cattle that Josiah had bought, and meeting with Ts'eh Montaño, "a surreal, semi-divine life force" (Snodgrass 216).

In addition to finding the speckled cattle and meeting Ts'eh, Tayo must also reconnect with the land and the people to be fully healed of the war trauma, and the familiar scents of the plants that he smells along the way play a vital role in this reconnection process. At one point in the novel, for example, he arrives at a canyon, where the smell coming from beeweed plants helps him to recognize the place: "The canyon was the way he always remembered; the beeweed plants made the air smell heavy and sweet like wild honey, and the bumblebees were buzzing around waxy yucca flowers" (*Ceremony* 41). It is thus important for Tayo to see, touch, smell, and work with the plants to be able to reconnect with the land and the people around him. In contrast to the dominant modern culture that views plants and the rest of the nature from a materialistic and utilitarian perspective, Tayo must understand an integral and spiritual relationship with the plants and animals. For this reason he "[steps] carefully, pushing the toe of his boot into the weeds first to make sure the grasshoppers were gone before he set his foot down into the crackling leathery stalks of dead sunflowers" (*Ceremony* 136). In Tayo's view, plants and animals are not merely objects to be used by the humans; rather, they possess life – "leathery stalks" – and should be treated with care. Contrast this view with that of Emo, who buys into the destructive modernist perspective in his treatment of nature and other humans: "He [Emo] went into the old man's field to look at the melons, all round and full of slippery sinews of wet seeds. He raised his foot carefully and brought his boot down hard on the center of the melon. It made a popping sound. Seeds and wet pulp squirted out from the broken rind; they glistened with juice. He kicked the pieces and scattered them around the corn plants" (*Ceremony* 56-57). Emo's violent attitude toward nature is indicative of the worldview that he shares, and it is such worldview that Tayo must resist in order to recover.

While nature as a whole plays a crucial role in Tayo's recovery, this essay focuses on the way plants help him in reconnecting with the land and the people. Plants in Native American culture, as Ken Cooper reminds us, are regarded as "standing people," and the Native Americans hold a friendly relationship with them (Cooper 116). In keeping with Cooper, I argue that plants function like an old friend for

Tayo, and without them, it would be very difficult for Tayo to fully come to his own; plants help him even when his human friends desert him. I make this argument against a critical backdrop *Ceremony* scholarship throughout which plants have been largely ignored in analyzing how Tayo is able to successfully fight against the "witchery" that has plagued his fellow veterans such as Emo, Harley, and Leroy Valdez.

In her essay, "Contested Ground: Nature, Narrative and Native American Identity in Leslie Marmon Silko's *Ceremony*," Rachel Stein interprets the novel as describing "Tayo's healing as he comes to use his cultural confusion and ambiguous social position in order to renegotiate the relations between Indian and white cultures through a ceremonial reconstruction of the traditional Laguna relationship to the land" (197). Stein correctly identifies the "fatal opposition between the aboriginal stories of partnership and reciprocity with nature and the Euro-American stories of detachment and dominion" to be the underlying reason for Tayo's illness, and its cure, Stein argues, lies in his successful untangling of "this conflict of paradigms" (196). Stein further argues that Tayo is able to transcend the conflicting paradigms by establishing "a living relationship to the Laguna sacred principles incarnate in nature," in which Ts'eh plays an important role (206). While Stein's analysis takes into account how dismantling fences erected between indigenous and modern culture is important in Tayo's recovery, and how Ts'eh as a hybrid figure teaches him to dismantle such fences, her essay does not devote much space to discussing how Ts'eh helps Tayo in reconnecting with the land and the people through the study of plants.

Similarly, in his essay "Writing Nature: Silko and Native Americans as Nature Writers," Lee Schweninger locates Tayo's recovery in "his ability to learn not to hate the destroyer, for in hating the destroyer he too becomes a destroyer" (55). He references an important passage in the novel in which Tayo reflects on finding and planting the seed of a flower that Ts'eh had instructed him to plant before parting with him the last time. It is worth reproducing the quote in full as it is central to understanding the role plants play in Tayo's recovery:

> He would go back there now, where she [Ts'eh] had shown him the plant. He would gather the seeds for her and plant them with great care in places near sandy hills. The rainwater would seep down gently and the delicate membranes would

> not be crushed or broken before the emergence of tiny fingers, roots, and leaves pressing out in all directions. The plants would grow there like the story, strong and translucent as the stars. (*Ceremony* 236)

Tayo ponders this right after he narrowly escapes Emo's assassination plot. As I will discuss in detail later in the paper, Tayo's ability to identify and plant a flower "with great care" indicates that he can now relate to his land; he is committed to follow through the instructions he has received from Ts'eh about plants. At one point, she tells him that she will take a special type of plant, and "plant it in another place, a canyon where it hasn't rained for a while" (*Ceremony* 208), suggesting that, among other things, plants play an important role in bringing rain, which is vital for both humans and animals. The fact that Tayo begins to perceive plants as a living entity – "the delicate membranes would not be crushed or broken before the emergence of tiny fingers" – is symptomatic of his ability to resist the temptation to be a destroyer like Emo. And yet, Schweninger leaves the passage above, which is the only important passage concerning plants cited in his essay, underanalyzed.

Edith Swan also briefly refers to the passage above in her essay, "Healing via Sunwise Cycle in 'Ceremony,'" and argues that Tayo refuses to participate in the cycle of violence perpetuated by modernity and its agents in the novel, such as Emo; instead "he plants the seeds of Ts'eh's flowers of 'light'" (Swan 325). However, she, like Schweninger, leaves unexplored Ts'eh's study and teaching of plants and its impact in Tayo's healing. While she certainly discusses how Betonie deploys trees and bushes during Tayo's healing ceremony, Swan devotes most of her essay to exploring Tayo's successful understanding and performance of Navajo ritual practices. She contends that Tayo "must continue to internalize aspects of 'mountain' into his nature" as per Betonie's teaching that "[i]t is people who belong to the mountain," not the other way around (Swan 322). It seems to me that while revering the mountain and the land constitutes an important aspect of Tayo's successful healing, he also must find a way to connect with them, and I argue that plants, along with animals, help him in this act.

Other critics have focused on either laughter or connection with language as a necessary part of Tayo's healing.[1] Silko herself has underscored Tayo's ability to connect with the rest of humanity as an important component of his healing. Talking with Per Seyersted, Silko remarks that "the Destroyers and the destructive impulse don't reside with a single group or a single race, and that to manipulate people into war or other conflicts again is a human trait; it is a worldwide thing" (Seyersted 36). And it is Tayo's ability, as Betonie tells him, to transcend the divisive understanding of humanity and to be able to see a complex dynamic as it relates to good and evil that will ultimately determine whether he will successfully recover. Silko further notes, "Tayo's healing is connected to the faith which this old medicine man [Betonie] had, a faith which went back to things far in the past, the belief that it's human beings, not particular tribes, not particular races or cultures, which will determine whether the human race survives" (Seyersted 35). I would add that Tayo must learn an interconnection that marks not only human relations but also the relation between humans and the rest of the nature. Plants play an important role in helping Tayo reconnect with the land and the peoples and in enabling him to perceive the manner in which the fate of humanity is interconnected with the fate of the earth.

In her essay, "Landscape, History, and Pueblo Imagination," Silko discusses how in American Indian culture the universe is regarded as a family unit, in which the Earth and the Sky are sisters. Harmony, she points out, marks all relations between humans and non-humans: "Survival depended [in ancient Pueblo culture] upon harmony and cooperation not only among human beings, but among all things – the animate and the less animate, since rocks and mountains were known to move, to travel occasionally" ("Landscape" 267). In the Pueblo imagination, every creature, plant, and rock occupies a special place,

---

[1] See for example Elizabeth N. Evasdaughter, "Leslie Marmon Silko's Ceremony: Healing Ethnic Hatred by Mixed-Breed Laughter," *MELUS* 15.1 (1988): 83-95. In the essay Evasdaughter discusses how Silko uses comicality to deal with tragedy brought about by the "witchery." She argues, "Human clowning of a farcical type, exposing human flaws in manifestly physical way, builds up Silko's philosophy" (85). Gloria Bird, on the other hand, focuses on language and how Tayo must learn to overcome a colonialist discourse that "serves to continue the paradigm of the native as Other by imbibing that Otherness within the tropes of moral and metaphysical differences" (3). See her essay, "Toward a Decolonization of Mind and Text 1: Leslie Marmon Silko's 'Ceremony,'" *Wicazo Sa Review* 9.2 (1993): 1-8.

because they help each other in the survival cycle: "Nothing is overlooked or taken for granted" (275). If any facet of creation, such as the landscape, is not accorded a proper respect or attention, it could prove fatal. Silko gives an example of a high dark mesa in Swanee, New Mexico where her great-grandmother's uncle and his brother-in-law lost their lives together with their herd of sheep for not paying attention to the place in which a group of Apache raiders were hiding. Survival depends, Silko further argues, in according proper care and respect to the landscape, "making the best use of all available resources" (271). Tayo not only respects all plants and animals and rocks, he also makes best use of them during his final fight against the "destroyers." As a result, he is able to detect just in time how the "witchery" has turned his friends into enemies: "Tayo was halfway to the top of the hill before he stopped; suddenly it hit him, in the belly, and spread to his chest in a single surge: he knew then that they [Leroy and Harley] were not his friends but had turned against him, and the knowledge left him hollow and dry inside, like the locust's shell" (*Ceremony* 225). He then hides among trees and rocks to save himself from Emo's plot to kill him.

Tayo makes spiritual contact with plants throughout the novel. Even prior to his deployment in the Pacific war, while Josiah was still alive, he felt happy smelling and touching plants:

> They had unloaded them [the Mexican cattle] on the Dedillo Grant because the grass was still good down there. Josiah wanted to give them a good start because they would be calving later on. […]
>
> They rode south with the sun climbing up in the east, making the sky bright, almost blinding. There were no clouds and the air still smelled cool. He wanted to remember the morning, bright and clear as the leaves on the little green plants which grew low and close to the sandy ground. It had the clarity of the sky after a summer rainstorm, when the dust was washed away, and the colors of the hills and the shadows of the mesas had an intensity which made everything he saw accessible, as if he could touch all of it, even the little green rabbit weed growing close to the sand, its tiny leaves clustered like stars. (*Ceremony* 72)

Not only does the green grass help Tayo and Josiah keep the cattle, who would otherwise have traveled south, settled down, but the other plants such as "the little green rabbit weed" put Tayo in contact with the land and the universe as he sees patterns of stars in the way the leaves of green rabbit weed cluster. In other words, the plants are important to not only feed the cattle, but also to connect with the rest of the universe through them. Tayo, in particular, likes the smell coming from "the little green plants" that makes him feel good, despite the misery he suffers at the hands of Auntie.

Tayo also associates certain plants with the people who constitute an important part of his life. The Night Swan, a Mexican dancer in Lalo's bar in Cubero, is one of these important people. Josiah visited her for love before his death, and Tayo makes love to her before departing for the war. After making love to Tayo, she tells him how people are afraid of change, and how they blame their misfortunes on people who look different. She then asks Tayo to remember that he is part of the human narrative, and as long as he is in it, he should not feel like an outsider: "'You don't have to understand what is happening. But remember this day. You will recognize it later. You are part of it now'" (*Ceremony* 92). Near Lalo's bar, where the Night Swan lived, there stood an old cottonwood tree, which Tayo had noticed during his tryst with her: "He walked up the spiral staircase [to the Night Swan's place], smelling wet adobe plaster and listening to the rain rattle the waxy green cottonwood leaves growing near the porch" (*Ceremony* 90). Now, the cottonwood tree is not just an old tree for Tayo, but an old friend.

After the war, he goes to Lalo's bar to see Night Swan's place, which has been deserted since Josiah's death. Tayo feels alive again seeing the cottonwood tree that stands just like it did when the Night Swan still lived there:

> He sat down on the upstairs porch with his back against the adobe wall and closed his eyes. *In a world of crickets and cottonwood trees he was almost alive again; he was visible.* The green waves of dead faces and the screams of the dying that had echoed in his head were buried. The sickness had receded into a shadow behind him, something he saw only out of the corners of his eyes, over his shoulder (*Ceremony* 95-96, emphasis mine).

Tayo feels better when he makes connection with his life before the war, and plants, the cottonwood tree in this particular instance, play an important role in making this connection.

The smell of "the waxy dark green leaves" of the cottonwood tree near the Night Swan's old place also takes Tayo back to his childhood days, when they used the cottonwood berries as ammunition against other boys as they played:

> He smelled the waxy dark green leaves, and remembered climbing the big cottonwood trees along the river and plucking heavy hanging bunches of cottonwood berries that grew on the female trees late in the summer. They had carried home armloads of the seed pods, the size of peas, full of fluffy cotton when they split open. The smell of the crushed leaves had been exciting then because the cottonwood berries were ammunition to use against the other boys from the village, who tucked their shirttails into their jeans and filled the inside of the shirts with cottonwood berries until they had pendulous bellies. (*Ceremony* 95)

Seeing and smelling the cottonwood tree and hearing the sound of crickets give Tayo energy to once again engage the world. Without the smell and the sound coming from plants and insects, it would be very hard for Tayo to reconnect with the land and the people, and ultimately to find meaning in life. However, seeing and smelling the tree reunites him with his life in the past when Rocky, Josiah, and the Night Swan were still around him; it is as if the cottonwood tree now represents all the people whom he has loved but who are no longer with him physically: Rocky, Josiah, the Night Swan, and his mother.

The cottonwood tree also reminds Tayo of his mother, who lived her brief life like the Night Swan. This is what he had heard about his libertine mother from Auntie: "'Right as the sun came up, she [Tayo's mother] walked under that big cottonwood tree, and I could see her clearly: she had no clothes on. Nothing. She was completely naked except for her high-heel shoes. She dropped her purse under that tree. Later on some kids found it there and brought it back. It was empty except for a lipstick" (*Ceremony* 65). The cottonwood tree, then, serves for Tayo several purposes: it reminds him of his old days with Uncle Josiah and Rocky, and of his childhood. It joins him with the Night Swan, and his mother, both of whom are now either dead or physically absent. Ultimately, it helps him reconnect with the mother earth. Speaking of the organic relationship between the human beings and the earth, Silko writes, "The dead become dust, and in this becoming they are once more joined with the Mother. The ancient Pueblo people called the earth the Mother Creator or all things in this world. Her sister, the Corn Mother, occasionally merges with her because all succulent green rises out of the depths of the earth" ("Landscape"

265). If the dead turn into dust, and the dust feeds the cottonwood tree, then, it makes perfect sense that Tayo feels well seeing and smelling the leaves of a cottonwood tree.

At one point in the novel, Tayo remembers the story Josiah had told him and Rocky about the interconnection between humans, animals, and plants. Josiah had told them that when the mother of the people got angry over the way people were misbehaving, "[the] animals disappeared, the plants disappeared, and no rain came for a long time. It was the greenbottle fly who went to her, asking forgiveness for the people. Since that time the people have been grateful for what the fly did for us" (*Ceremony* 93). This connection that Josiah had made between animals, plants, and humans guides Tayo's course in the present, as is evident in the passage quoted at the beginning of the paper in which Tayo carefully treads on the dead sunflower plants lest he step on the grasshoppers. Similarly, when he gathers pollen from flowers, he displays equal care in making contact with the plants: "He found flowers that had no bees, and gathered yellow pollen gently with a small blue feather from Josiah's pouch; he imitated the gentleness of the bees as they brushed their sticky-haired feet and bellies softly against the flowers" (204-5). Here, as elsewhere, Tayo treats the flowers and animals with care, something he learned from Josiah before the war, and from Betonie and Ts'eh afterwards.

After the war, the most important person, besides Betonie, who helps him find himself is Ts'eh. Writing about this elusive figure, Patricia Clark Smith and Paula Gunn Allen note, "She is a mountain spirit, like her brothers and sisters – sacred mountains all. Though she and Tayo are lovers, in scenes among the most erotic in American literature, her sexuality extends far beyond the act of intercourse – she is healer, nurturer, plotter, planter, and she schemes for the good of people and plants and animals" (Smith with Allen 140). Not surprisingly, this planter incarnate Ts'eh is standing under an apricot tree at the time Tayo first meets her while he is searching for Josiah's cattle (*Ceremony* 164). She then invites him into her house as if she had been waiting for his arrival all along: "He followed her down the long screen porch to a narrow pine door. When she opened it, he smelled dried apricots and juniper wood burning. The inner walls were massive and all the doorways were low. The smell of clay and mountain sage stirred old memories" (165).[2] As the last sentence of the quote hints, Ts'eh reminds Tayo of the Night Swan, as he had smelled a similar scent when he walked up the staircase to the Night Swan's place to deliver Josiah's note.

Once inside, Tayo immediately feels at home; the mud plastered walls in the house look to him as timeless as the mesas surrounding the house. Then Tayo observes that the house is full of plants both inside and out:

> Along the south wall, tall orange sunflowers were still blooming among dry corn stalks; the wind of the night before had twisted the sunflowers around the brittle corn stalks, so that in the early morning light the dried-up corn plants were bearing big orange sunflowers that dusted the hard-packed earth beneath them with orange pollen. Somebody had planted blue morning-glories below each of the four wide windows, and the vines of the blue flowers were climbing cotton strings that had been nailed to the window frames. The morning-glories were open wide, themselves the color of the sky, with thin white clouds spreading from the center of the blossoms into the bright blue. (*Ceremony* 170)

Ts'eh the planter has surrounded her house with different types of plants, for each plant bears special meaning for her, as Tayo learns from her later. As Tayo sits in the house savoring its details, she works with rocks and plants: "She reached into a flour sack by her feet and brought out bundles of freshly gathered plants. She sniffed them and blew on them before she matched the plants with the stones, putting a sprig of blue-gray mountain sage with the blue stone" (170). She matches the color of the plant with the color of the stone to determine what meaning each plant carries, and where it can be found. As Silko reminds us elsewhere, Native Americans believe that the stone possesses spirit or being just like plants and animals ("Landscape" 265), and that if we recognize this potency, as Ts'eh does, it can prove tremendously useful in solving human problems such as drought.

When Tayo volunteers to shake off snows from the apricot tree in the front yard of Ts'eh's house, he remembers the time he and Rocky

---

[2] Apricot is also mentioned in relation with the "Yellow Woman" in Silko's short story of the same title, suggesting an association between the apricot tree and female sexuality. In fact, Mary Ellen Snodgrass argues that the "stone fruit on the inside mirrors the outline of the female vulva and vagina" (Snodgrass 215).

had shaken the budded apple trees: "'I am going out. I'll shake the snow off the branches,' Tayo said, remembering how one spring when a late snow fell he had helped Josiah and Rocky shake the budded apple trees" (*Ceremony* 193). Then, following a ritualistic circle, he shakes off the snow from the branches: "He shook the snow off carefully, moving around the tree from the east to the south, and from the west to the north, his breath steaming out in front of him" (194). Tayo forms a circle by moving from north to east to west and back to north. As Paula Gunn Allen notes, in Native American culture, the circle or "sacred hoop" represents unity "that is dynamic and encompassing, including all that is contained in its most essential aspect, that of life" (Allen 243). Tayo's acknowledgement of the interdependence between him and the tree is demonstrated in the way he takes care when touching the tree so that he does not do to it more harm than good.

Tayo has started to show a full recovery of his health as he comes to feel that Josiah and Rocky are not very far from him as long as he hears the buzzing of grasshoppers wings coming from the weeds in the yard, despite the destruction of the white ranchers, who log the trees, kill the animals for sport, and build fences. He reflects how the land and the people are interconnected, and how the destroyers cannot separate them: "The mountain could not be lost to them, because it was in their bones; Josiah and Rocky were not far away. They were close; they had always been close. And he loved them then as he had always loved them, the feeling pulsing over him as strong as it had ever been" (*Ceremony* 204). The spiritual presence of his cousin and uncle around him helps him stop blaming himself for their deaths, showing signs of recovery.

When Ts'eh reappears to him in the ranch, she "was walking through the sunflowers, holding the blue silk shawl around her shoulders in one hand, carrying the long curved willow stick in the other" (*Ceremony* 205). The willow tree functions here, as in Silko's short story "Yellow Woman," as a plant that connects and reconnects people. In "Yellow Woman," the titular character meets her elusive lover, Silva, when the latter was "sitting on the river bank trimming leaves from a willow twig with his knife" ("Yellow Woman" 35). In *Ceremony*, Tayo meets again with his own elusive lover, from whom Tayo learns more about different plants:

> He went with her [Ts'eh] to learn about the roots and plants she had gathered. When she found a place she got comfortable, spreading her blue shawl on the ground after she had cleared the area of pebbles and little sticks and made sure no ants were disturbed. She sat flat on the ground and bent over close to the plants, examining them for a long time, from the petals, sprinkled with pollen, down the stem to each leaf, and finally to the base, where she carefully dug the sand away from the roots. (*Ceremony* 208)

Ts'eh, like Tayo, takes care not to disturb "the animal people," the ants, while she uproots a plant. She then points to a plant that has a color of the sky, and tells him that the plant would bring rain when she plants it in a canyon where it has not rained for a while (209).

Ts'eh asks Tayo to collect the seeds of a plant for her when it is ripe, obviously in an attempt to allow him to practice his newly acquired knowledge about plants; the request could also be read as Ts'eh's attempt to continue her relationship with Tayo to make sure that he is fully connected to the land by way of connecting with the plants:[3] "'This one,' she said, pointing at a tall dark green plant with round pointed leaves, deep veined like fossil shells. The flat seed pods were still thick and green, but later, in the fall, the skin would dry thin, and cold winds would strip away the hull to the last translucent membrane, holding the dark eyes of the seed inside it" (*Ceremony* 210-11). Ts'eh tells Tayo that the plant represents the "light of the stars, and the moon penetrating the night" and that if care is not taken while handling it, the stem would break, and "its wet flowing vitality would be lost in a single breath" (211). In line with Ken Cooper's designation of plants as "standing people" (Cooper 116), Ts'eh teaches Tayo to re-

[3] The relationship to the land, as Paula Gunn Allen reminds us, is an important part of Native American identity. Allen asserts, "We are the land. To the best of my understanding, that is the fundamental idea embedded in Native American life and culture" (qtd. in Corbett 50). Allen also clarifies how Native American people relate to plants as equal fellow beings:

In the Native American system, there is no idea that nature is somewhere over there while man is over here, nor that there is a great hierarchical ladder of being on which ground and trees occupy a very low rung, animals a slightly higher one, and man a very high one indeed. ...All are seen to be brothers or relatives..., all are offspring of the Great Mystery, children of our mother, and necessary parts of an ordered, balanced, and living whole. (qtd. in Corbett 51)

member that there is life-blood flowing inside the vein of the plant just like it does in humans and animals.[4]

When Tayo successfully fends off the destructive "witchery" of Emo, who has totally abandoned himself to the temptations of modern life, Tayo remembers the job placed on him by Ts'eh. It is about time that the plant is ready to yield seeds, which he hopes to plant in places where it has not rained for a while. The rainwater would then help the seed grow shoots like tiny fingers in all directions. He hopes that the plants emerging from the seeds "would grow there like the story, strong and translucent as the stars" (*Ceremony* 236). Plants constitute part of the cosmic story, and his job is to join hands with them.

However, not all plants that grow all over the planet carry the same significance to everyone. To be able to find meaning in plants, one has to grow with them. Unfamiliar plants, just like unfamiliar people and animals, might strangle and suffocate us, a reality that Tayo has learned through his war experience in the Philippines: "The valley was green, from the yellow sandstone mesas in the northwest to the black lava hills to the south. But it was not the green color of the jungles [in the Philippines], suffocating and strangling the earth. The new growth covered the earth lightly, each blade of grass, each leaf and stem with space between as if planted by a thin summer wind" (*Ceremony* 203). Of course, people who have lived among the jungles might feel differently about them. In this context, Sheila Kelly explains that, "We interact with the people and the plants and the geography of our setting" (Kelly 106). Back home in his own "setting," which is different from that of the Philippines, Tayo reconnects with his setting organically with a little help from Betonie and Ts'eh: "The sun felt good; he could smell the juniper and piñon still damp from the rain. The wind carried a wild honey smell from meadows of beeweed" (*Ceremony* 205).

---

[4] Contrast the Native Americans attitude towards plants as "standing people" with the modernity's attitude toward nature as "standing reserve." Explaining the term that Martin Heidegger uses to describe the age of technology, Cynthia Deitering writes that "the Western cultural perception of nature and material objects was that of 'standing reserve' whereby a tract of land was revealed and represented as a coal mining district, a mineral deposit; or a river was regarded and represented as a supplier of water power [...]" (199). See her essay, "The Postnatural Novel: Toxic Consciousness in Fiction of the 1980s" in *The Ecocriticism Reader: Landmarks in Literary Ecology*, edited by Cheryll Glotfelty and Harold Fromm, (Athens, Georgia: University of Georgia Press, 1996), pp. 196-203.

In fact, Tayo grows up among juniper and piñon, apple and cottonwood trees, tamaric (or tamarisk) and willow in the yellow mesas. These plants not only help him now to connect back to the land after returning from the war, but also protected him while he was growing up. Tayo remembers a time when the police raided the slum in the arroyo where his mother lived with him. When the police caught everyone and hauled them to the van, he could hide in the tamarisks:

> He hid in the tamarics, breathing hard, his heart pounding, smelling the shit on his bare feet. The summer heat descended as the sun went higher in the sky, and he watched them [the police], lying flat on his belly in the dry leaves of tamaric that began to make him itch, and he moved cautiously to scratch his arm and his neck. (*Ceremony* 103).

After the war, he hides behind the cottonwood and juniper trees when Emo tries to assassinate him: "He was smiling and suddenly close to tears because they [Leroy and Harley] had come when he needed friends most. He stepped out from behind the juniper tree and waived both arms above his head" (*Ceremony* 221). Little does he know that even his close friends have been bewitched by Emo's "witchery" and are now set against him.

When his human friends turn against him, Tayo counts on his nonhuman friends such as the trees and the rocks. At the final moment of Emo's assassination plot, he hides between two boulders: "He crouched between the boulders and laid his against the rock to look up at the sky" (*Ceremony* 235). Having narrowly escaped the assassination plot, Tayo thinks now to go and gather the seeds of the plant that Ts'eh had entrusted him to do. Even though the destroyer Emo still lives, and the witchery continues to wreak havoc, he feels good to be alive in the company of the stars and plants and mountains.

In the movie *Foxfire* (1987), the central character, Annie Nations (played by Jessica Tandy), lives in the Blue Ridge Mountains, in a house built by her husband Hector, who has been dead for five years. Her country singer son, Dylan (played by John Denver), lives in Florida with his family and visits her from time to time to see if she has changed her mind to sell the property and move to Florida. He himself has changed since moving away to the city; only recently his wife has left him and the children for another man. Like his mother, he loves the place where he grew up, but he cannot move back to the mountain house because, like everyone else, he is lured by modernity. Moreover, he has children to take care of, who have grown in the city. However, in one of his visits to his childhood village to perform in a concert, he spends the night in the old house, where his mother still lives together with the spirit of her husband. The following morning, he comes out of the house holding a cup of coffee, looks at the green trees in the mountains and says aloud to himself: "Nice smell." When his mother asks if he was talking about the coffee, he says: "Not the coffee; this place." I think this is an example of "spiritual communion" with nature that Sheila Kelly talks about (Kelly 108). Tayo feels the same way as Dylan when he smells the plants in the mountains and mesas after coming home from the war. Unlike Dylan, however, Tayo decides to stay with the mountains and the trees. They mean a lot more to him than anything else.

It seems fit to conclude with a few lines from Silko's poem, "COTTONWOOD Part One: Story of the Sun House:"

> Cottonwood,
> cottonwood.
>
> So much depends
> upon one in the great canyon.[5]

Works Cited

Allen, Paula Gunn. "The Sacred Hoop: A Contemporary Perspective." *The Ecocriticism Reader: Landmarks in Literary Ecology*. Ed. Cheryll Glotfelty and Harold Fromm. Athens, Georgia: University of Georgia Press, 1996. 241-263. Print.

Bird, Gloria. "Towards Decolonizing the Mind and Text 1: Leslie Marmon Silko's 'Ceremony.'" *Wicazo Sa Review* 9.2 (1993): 1-8. Print.

---

[5] The full poem, which has two parts, is included in the collection, *Storyteller* by Leslie Marmon Silko.

Cooper, Ken. "A Native Elder Speaks: How Do You Say Good-bye to a Friendship?" *Earth and Spirit: The Spiritual Dimension of the Environmental Crisis*. Ed. Fritz Hull. New York: Continuum, 1993. 116-118. Print.

Corbett, Julia B. *Communicating Nature: How We Create and Understand Environmental Messages*. Washington: Island Press, 2006. Print.

Evasdaughter, Elizabeth N. "Leslie Marmon Silko's Ceremony: Healing Ethnic Hatred by Mix-Breed Laughter." *MELUS* 15.1 (1988): 83-95. Print.

Foxfire. Dir. Jud Taylor. Perf. Jessica Tandy, Hume Cronyn and John Denver. Hallmark Hall of Fame, 1987. DVD.

Kelly, Sheila. "The Path of Place." *Earth and Spirit: The Spiritual Dimension of the Environmental Crisis*. Ed. Fritz Hull. 105-113. Print.

McMurtry, Larry. "Introduction." *Ceremony*. New York: Penguin Books, 2006. xxi-xxiii. Print.

Schweninger, Lee. "Writing Nature: Silko and Native Americans as Nature Writers." *MELUS* 18.2 (1993): 47-60. Print.

Seyersted, Per. "Interview with Leslie Marmon Silko." *Conversations with Leslie Marmon Silko*. Ed. Ellen L. Arnold. Jackson: University Press of Mississippi, 2000. 29-36. Print.

Silko, Leslie Marmon. *Ceremony*. 1977. New York: Penguin Books, 2006. Print.

—. "Landscape, History and Pueblo Imagination." *The Ecocriticism Reader: Landmarks in Literary Ecology*. Ed. Cheryll Glotfelty and Harold Fromm. 264-276. Print.

—. *Storyteller*. New York: Arcade Publishing, 1981. Print.

—. "Yellow Woman." *"Yellow Woman."* Ed. Melody Graulich. New Brunswick, NJ: Rutgers UP, 1993. 31-43. Print.

Smith, Patricia Clark with Paula Gunn Allen. "Earthy Relations, Carnal Knowledge: Southwestern American Indian Women Writers and Landscape." *"Yellow Woman."* Ed. Melody Graulich: 115-150. Print.

Snodgrass, Mary Ellen. *Leslie Marmon Silko: A Literary Companion*. Jefferson, North Carolina: McFarland & Company, 2011. Print.

Stein, Rachel. "Contested Ground: Nature, Narrative, and Native American Identity in Leslie Marmon Silko's *Ceremony*." *Leslie*

Marmon Silko's Ceremony: A Casebook. Ed. Allan Chavkin. New York: Oxford UP, 2002. 193-211. Print.

Swan, Edith. "Healing via the Sunwise Cycle in Silko's 'Ceremony.'" *American Indian Quarterly* 12.4 (1988): 313-328. Print.

The Bible's Paradise and *Oryx and Crake*'s Paradice: A Comparison of the Relationships between Humans and Nature

Rhona Trauvitch

Abstract: This paper explores the relationship between humans and plant life as depicted in two creation stories: that in the Bible, and that in Margaret Atwood's novel, *Oryx and Crake*. An analysis of the relationship between Genesis' first two humans and nature is revealing in terms of Crake's possible motivation when creating the Crakers. Specifically, it appears that Crake's purpose may be to re-access a state of pre-fall Paradise. A comparison of the two creation stories foregrounds the question of whether human qualities - the characteristics that separate humans from other creatures - might be at odds with a harmonious state of nature.

In her 2003 novel *Oryx and Crake* Margaret Atwood invokes the biblical story of Genesis in several ways. The primary and most direct conjuring stems from the MaddAddam group and their game of Extinctathon. Upon logging on to the game's website, one reads "*EXTINCTATHON, Monitored by MaddAddam. Adam named the living animals, MaddAddam names the dead ones. Do you want to play?*" (80). This references Genesis 2:19 and 2:20, in which God brings the animals to Adam so that he will name them. Atwood further invokes Genesis with God's Gardeners and their religion – briefly mentioned several times in *Oryx and Crake* and taking a central role in the 2009 follow-up, *The Year of the Flood.* Genesis is also implicated in Crake's attempts to recreate Paradi[c]e with a new set of inhabitants: a genetically engineered human species referred to as the Crakers.

A comparison of the creation story in Genesis to that in *Oryx and Crake* – specifically, a comparison of the relationship between man and the environment that is depicted in each – affords us a clearer view of the purpose, design, and potential of the Crakers. It suggests that what Crake attempts to do in his design of the Crakers is akin to a rewinding of history to a point analogous to Paradise before the fall. Crake tries to accomplish the rewinding by releasing a pathogen that is meant to eradicate from Earth all humans but the Crakers. Moreover, Crake resets the parameters of nature by changing the qualities of "human nature" in the Crakers' design. Upon a mere rewinding, history would ostensibly repeat itself. Crake has to both rewind and reset.

An analysis of man's relationship to his environment – the surrounding plant life – in each story indicates that Crake might question whether human qualities are at odds with equilibrium in nature. By ascertaining man's relationship to the plant life that surrounds him in each of the two scenarios, we can determine a corresponding degree to which he exhibits human qualities: the qualities that differentiate him from the other creatures in his kingdom of classification. The qualities in question are not genetic differences – markers or traits that taxonomically separate the *Homo sapiens*. Rather, the qualities in question are those that pertain to capacities for complex reasoning, agency, and theory of mind. The beings that Crake forms are – whether for the purpose of survival or as a result of his own whims – designed to be similar to man before the fall and to remain in that state perpetually.

Plants in the Creation Story: Genesis

In my analysis of biblical passages, I rely for the most part on two sets of commentary: that by Rashi (1040 – 1104), and that by Rabbi Elie Munk (1900-1981). Rashi is the acronym by which Rabbi Shlomo Yitzhaki is known. Rashi is perhaps the most prominent, canonical, and widely read of biblical commentators and his insight is indispensible to our analysis of biblical verse. Munk's perspective is likewise especially useful: In *The Call of the Torah*, Munk offers his own astute perspectives on the Bible in conversation with the scores of commentators he references.

Genesis 1-3 tells the story of creation and culminates in man's fall from Paradise. The relationship between man and the environment is clarified when we read the story with an eye to the depiction of vegetation – when and how it is mentioned, and the manner with which it is to be treated. The first mention of vegetation is in Genesis 1:11 and 1:12, which convey that vegetation came to be on the third day of creation:

> And God said, "Let the earth sprout vegetation, seed yielding herbs and fruit trees producing fruit according to its kind in which its seed is found, on the earth," and it was so. And the earth gave forth vegetation, seed yielding herbs according to its kind, and trees producing fruit, in which its seed is found, according to its kind, and God saw that it was good.[1]

On the sixth day, upon the creation of man and woman, God reveals the type of relationship that he intends for there to be between humans and their environment:

> And God said, "Behold, I have given you every seed bearing herb, which is upon the surface of the entire earth, and every tree that has seed bearing fruit; it will be yours for food. And to all the beasts of the earth and to all the fowl of the heavens, and to everything that moves upon the earth, in which there is a living spirit, every green herb to eat," and it was so. (Gen. 1:29–1:30)

We can deduce from the order of creation and from what God tells man that vegetation is available for human sustenance. In Genesis 2, we find out more about the relationship between man and vegetation:

> Now no tree of the field was yet on the earth, neither did any herb of the field yet grow, because the Lord God had not brought rain upon the earth, and there was no man to work the soil. And a mist ascended from the earth and watered the entire surface of the ground. And the Lord God formed man of dust from the ground [...] And the Lord God planted a garden in Eden from the east, and He placed there the man whom He had formed. And the Lord God caused to sprout from the ground every tree pleasant to see and good to eat, and the Tree of Life in the midst of the garden, and the Tree of Knowledge of good and evil. (Gen. 2:5–2:9)

Rashi comments on Genesis 2:5 as follows:

> [T]hey [the plants] had not yet emerged, but they stood at the entrance of the ground until the sixth day. And why? [...] Because there was no man to work the soil, and no one recognized the benefit of rain, but when man came and understood that they were essential to the world, he prayed for them, and they fell, and the trees and the herbs sprouted.

[1] While my analysis of the Biblical texts is done in the original Hebrew, all translations of the Bible in this paper, as well as of Rashi's commentary, are from Chabad.org, copyright The Judaica Press.

Here Rashi is referring to a section in the Halakhah, which states:

> R. Assi pointed out a contradiction [between verses]. One verse says: And the earth brought forth grass, referring to the third day, whereas another verse when speaking of the sixth day says: No shrub of the field was yet in the earth. This teaches us that the plants commenced to grow but stopped just as they were about to break through the soil, until Adam came and prayed for rain for them; and when rain fell they sprouted forth. This teaches you that the Holy One, blessed be He, longs for the prayers of the righteous. (Chul. 60b)

This creative reading of Genesis 2:5 implies that while the potential for vegetation was existent, its growth hinges on the active participation of man. Munk adds to the above interpretation Recanati's conclusion, "it follows that the benediction which comes from Heaven, to "water the ground everywhere" and to make the earth fertile, depends on man, on his prayer, on his active will" (Munk 26). Munk continues,

> Just as the work of the six days of creation reaches completion, the Torah makes us see that from now on, man is to play a key role in developing this creation. He is to be God's collaborator; he is charged with keeping the harmony of the universe. Everything will depend on his acts, on his conduct, and on his efforts. (26-27)

The choice of words – 'key role,' 'collaborator,' 'charged with,' 'depend on' – is indicative of a very high level of participation on the part of man. Considered together, these commentaries from Rashi, the Halakhah, Recanati, and Munk reveal something quite fascinating: in the biblical story of creation man's role as far as vegetation is concerned is that of an associate. Man's active will draws the plants from the seeds that are waiting at the gate of the ground. Man "activates" the plants via his collaboration.

Contrasted to this is man's role with relation to the animals. God grants man dominion over both nature and animals: "And God blessed them, and God said to them, "Be fruitful and multiply and fill the earth and subdue it, and rule over the fish of the sea and over the fowl of the sky and over all the beasts that tread upon the earth"" (Gen. 1:28). However, within this dominion, man's relationship to each is quite different. While in terms of plant life man is collaborator, in terms of the animals he is the characterizer and classifier. In Genesis 2:19 and 2:20 Adam names the animals:

> And the Lord God formed from the earth every beast of the field and every fowl of the heavens, and He brought [it] to man to see what he would call it, and whatever the man called each living thing, that was its name. And man named all the cattle and the fowl of the heavens and all the beasts of the field, but for man, he did not find a helpmate opposite him.

Munk comments,

> Adam characterized and classified the animals when he gave them their names [...] This name-giving required profound wisdom on Adam's part, for the name had to correspond to the nature, species, and sex of each creature [...] man possesses the faculty of recognizing the nature of things and of giving them names [...] the Torah stresses that the name given to each animal by man – הוא שמו – is a valid, definitive name. (38-39)

This interpretation is in line with the kabbalistic notion that Hebrew is a holy language, and that the letters of the Hebrew alphabet are far from arbitrary: they are in fact the building blocks of creation. According to this idea the name of an object corresponds to the object's nature or essence. In his overview to Rabbi Michael L. Munk's book *The Wisdom in the Hebrew Alphabet*, Rabbi Nosson Scherman describes the power of the Hebrew letters as follows: "The twenty-two sacred letters are profound, primal spiritual forces. They are, in effect, the raw material of Creation. When God combined them into words, phrases, commands, they brought about Creation, translating His will into reality, as it were" (19).

We can see this view of the Hebrew letters depicted in several ancient Jewish texts and legends. In his collection *The Tree of Souls: The Mythology of Judaism*, Howard Schwartz includes a section entitled "Myths of the Holy Word." Recounting the various myths in the section, Schwartz relates:

> God looked into the Torah and created the world and all created beings through it [...] God consulted only the Torah, and let the Torah serve as a blueprint for all creation. So too did the Torah serve as an artisan in all the work of creation [...] not only was the Torah created prior to the creation of the world, it was the vessel

> by which the world was created. Thus the universe was created through the letters of the Torah [...] With letters, heaven and earth were created, the oceans and rivers were created, all the world's needs and all the orders of creation. (249, 251)[2]

We thereby have a better understanding of why Munk emphasizes Adam's faculty of recognition and his profound wisdom. To name each animal Adam had to have been able to completely comprehend the animal's nature. Scherman writes, "Once we have achieved a rudimentary understanding of the spiritual content of the Hebrew letters [...] Adam's task takes on major dimensions [...] this man was truly being asked by God to demonstrate a spiritual insight that was profound beyond our imagination" (20). He continues with an example:

> When Adam said that a bull should be called שור and an eagle should be called נשר, he was saying that the spiritual forces expressed by those letters, in the formula signified by those unique arrangements of letters and vowels, were translated by God into the nerve, sinew, skin, size, shape, strength, and ability that we see when a sturdy bull pulls a plow or a soaring eagle excites our imagination." (20).

In naming – and therefore characterizing and classifying – each animal, Adam participates in nature in a manner different from the aforementioned collaboration with regard to vegetation. While in both cases there is participation, the nature and duration thereof is different. There is a finality in the naming of the animals: "and whatever the man called each living thing, that was its name." Rashi comments on 2:19, "Every living creature to which man would give a name-that was to be its name forever." Sherman writes, "The spiritual forces that Adam identified in the form of a name remain active for all time. The same spiritual forces that God translated into a שור at the dawn of creation remain the essence of a שור for all time" (20). The essence of the animal is recognized, the animal is classified, and that part is completed.

On the other hand, man's active participation in plant life is ongoing. As mentioned above, Munk notes that "from now on" man has a "role in developing this creation," and man's charge of "keeping the

[2] Schwartz provides extensive comments and citations after each myth, and so the interested reader can trace the myths back to their sources in rabbinic commentary kabbalistic texts, and ancient parables. For the sake of clarity and simplicity, I am limiting my own citations to Schwartz's formulations of the myths.

harmony of the universe." This is echoed later on when we read that man must tend to the vegetation: "Now the Lord God took the man, and He placed him in the Garden of Eden to work it and to guard it" (Gen. 2:15). Man's collaboration in this case is therefore clearly continuous.

Aside from – but related to – this difference between final and continuous, we have a difference in the nature of each participatory act. Naming (recognizing and classifying) is active, but far less active than developing, tending, and keeping. The act of naming calls on the faculties of reasoning and intellect, and is therefore more cerebral than physical, while the act of tending is more physical than cerebral: the nature of each act requires a different balance in the engagement of physical and cerebral faculties.

The relationship between man and his environment changes drastically after the fall. Munk compares the state of Adam's soul before the sin and after the sin in his interpretation of the role of the Tree of the Knowledge of Good and Evil:

> The tree which wakens in man the knowledge of good and evil. Right from creation Adam possessed free will, but as long as he did not taste the forbidden fruit his soul was suffused with an eternal inner glow [...]
>
> Only after Adam had tasted the forbidden fruit was desire kindled within him. Then forces of evil were roused in him and, along with them, instincts and sexual passions. The first sin destroyed the marvelous harmony of the beginning. (31)

Whereas before the fall Adam and Eve are in harmony with their surroundings in Paradise, after the fall they must toil both physically and spiritually to restore equilibrium. Munk writes, "We learn from the story of Paradise that the earth holds out to man the possibility of perfect harmonious co-existence with the elements of nature and with the animals" (48). Since the sin, however, everything man does is geared toward reaccessing Paradise and this harmonious co-existence with his environment. "The supreme aim of the Torah will henceforth be to have the gates of Paradise re-open and to restore the harmony which reigned on earth between the animal kingdom and mankind" (Munk 48-49).

Among the punishments that God delivers to man as a result of man's sin is the following pronouncement: "Because you listened to your wife, and you ate from the tree from which I commanded you saying, 'You shall not eat of it,' cursed be the ground for your sake; with toil shall you eat of it all the days of your life" (3:17). The sin therefore changes the very nature of the relationship between man and his environment. Munk interprets this verse as follows: "The fall of Adam entails that of all of nature [...] This is the important lesson of the unity of creation which this verse teaches: everything is closely connected. Everything is interdependent" (54). This notion of interconnection places an even greater emphasis on man's participation in the development of his surroundings. Munk notes that an improvement of man's situation now, after the fall, is possible but demanding: "the task has become more complex for it is much more difficult to restore an equilibrium that has been disturbed than to maintain one in its initial state. Henceforth, the whole problem of existence is to restore the harmony which originally existed" (55).

Having changed the essence of his nature by eating of the forbidden fruit, man is now faced with needs that were previously nonexistent. He has to develop tools so as to clothe himself, work the land, and build his shelter. He must now deal with pain and suffering. The break in the equilibrium that existed in Paradise creates the first sparks in what eventually become the fires of man's problems in the world: personal, physical, economic, social, and political.

Plants in the Creation Story: *Oryx and Crake*

Not unlike the attempts of linguists such as John Wilkins to recover the pre-Babel Adamic language, Crake attempts to reaccess Edenic conditions by re-creating/designing the human being in his own Paradice. There are two sets of humans in *Oryx and Crake*, and therefore two sets of relationships between people and nature: the relationship between the original humans and nature, and the relationship between the Crakers (humans 2.0) and nature. Regarding the first relationship, if we see present-day life as a continuation of the Bible's creation story, the humans in the setting Atwood describes have collaborated in creation to the extreme. Both forms of participation – that with animals and that with vegetation – have overtaken what is described and prescribed in Genesis. In terms of animals, man has ventured well beyond naming: he has mixed and spliced, thereby altering the very essences that he formerly characterized and classified. In terms of vegetation, man has moved beyond prayer and development: he has mixed and spliced within this kingdom as well, designing hybrid plants. And in both cases, partly in designing these new forms of life, man has also destroyed.

The animal splices are more apparent in the novel, what with the wolvogs, pigoons, and rakunks running around and often threatening humans, but the plant splices are substantial as well. There are the gaudy plants created by students in "Botanical Transgenics (Ornamental Division)": "a whole array of drought-and-flood-resistant tropical blends, with flowers or leaves in lurid shades of chrome yellow and brilliant flame red and phosphorescent blue and neon purple" (199). There are also the zucchinis that produce cheese (208). When Crake shows Jimmy around, presenting all the different splices, inventions, and genetically engineered innovations, Jimmy (understandably) expects "Some gruesome new food substance [...] A liver tree, a sausage vine. Or some sort of zucchini that grew wool" (302).[3] Even though none of these have been developed (yet), they are likely products because of the splices that we know have already been formed.

Perhaps the most prominent and environmentally, socially, and politically damaging creation in the vegetable kingdom is the Happicuppa bean.

> Until then the individual coffee beans on each bush had ripened at different times and had needed to be handpicked and processed and shipped in small quantities, but the Happicuppa coffee bush was designed so that all of its beans would ripen simultaneously, and coffee could be grown on huge plantations and harvested with machines. This threw the small growers out of business and reduced both them and their labourers to starvation-level poverty. The resistance movement was global. (Atwood 179)

In the case of the Happicuppa bean, the genetically engineered vegetation has led to some of the most horrific situations man has inflicted upon himself and the environment. We see in Atwood's apocalyptic novel the worst-case outcome of man's extreme meddling with his

[3] This is especially interesting because here we have plants that grow organs or flesh. It is a combination of the animal and vegetable kingdoms. We see that pigoons grow human organs, and encounter chickens that are grown without most of their brains and are therefore closer to vegetables, but the concept of a plant growing organs or flesh is a different one altogether. Rather than splicing within kingdoms, the engineers would be splicing kingdom with kingdom.

surroundings: a collaboration so exaggerated that it leads not to harmony with creation, but rather to the destruction thereof.

An analysis of the second aforementioned relationship – that between the Crakers and their environment – reveals that their design is an attempt to rewind and a reset the damage inflicted by humans and their extreme collaboration. The Crakers originate as altered human embryos, and are *sui generis* (303). They are programmed to die at the age of thirty, without sickness, thereby doing away with the issues of old age. They have "UV-resistant skin, a built-in insect repellant," and "an unprecedented ability to digest unrefined plant material" (304). They have an innate immunity from microbes. In terms of the makeover of the Craker's brain,

> Gone were its destructive features, the features responsible for the world's current illnesses. For instance, racism [...] had been eliminated [...] the Paradice people simply did not register skin colour. Hierarchy could not exist among them, because they lacked the neural complexes that would have created it. Since they were neither hunters nor agriculturalists hungry for land, there was no territoriality: the king-of-the-castle hard-wiring that had plagued humanity had, in them, been unwired. They ate nothing but leaves and grass and roots and a berry or two; thus their foods were plentiful and always available. Their sexuality was not a constant torment to them, not a cloud of turbulent hormones: they came into heat at regular intervals, as did most mammals other than man.
>
> In fact, as there would never be anything for these people to inherit, there would be no family trees, no marriages, and no divorces. They were perfectly adjusted to their habitat, so they would never have to create houses or tools or weapons, or, for that matter, clothing. They would have no need to invent any harmful symbolisms, such as kingdoms, icons, gods, or money. (305)[4]

Before discussing the implications of the Crakers' design, let us note several key differences between the creation of Adam and Eve and the design of the Crakers. The primary difference is the manner in

[4] An interesting consequence of the Crakers' design is that they would have no need for the genetically engineered vegetation that their makers – humans 1.0 – created. Many of the novel's genetically engineered plants were designed so as to profit corporations and cater to greed, the need of and propensity to which Crake has genetically engineered *out* of the Crakers. They will have no hierarchies and cannot conceive of profit. Moreover, the Crakers will have no need for the new plants as they are designed to feed on the "old" plants. This leads to curious pairings: the new plants correspond to the old humans; the new humans correspond to the old plants.

which the creatures are formed. Crake's creation is a second-tier one: God creates man (the progeny of Crake) and Crake created the Crakers. The first is a creation is *ex nihilo*[5] and the second iteration is in fact not really a creation at all: Crake does not *create* life – he mixes, splices, and engineers existing life into a new breed.

Further, Crake does not develop a new breed of vegetation for with the new breed of human: he places the new breed of human among existing vegetation. We see from the description of their design that no collaboration is asked of the Crakers. They are made to be compatible with the vegetation as is: "They are perfectly adjusted to their habitat." They are not required to pray, to actively develop vegetation, or tend to and work the soil. The Crakers are designed *according to* the vegetation that is already around them. Crake thus put into place, along with these new humans, an entirely new set of stipulations. The Crakers are designed for co-habitation and compatibility – not active collaboration.

These differences in formation point to the fact that Crake's is not a redoing of human life on Earth, but a rewinding thereof. After rewinding, Crake also resets the parameters of human interaction with the environment, and it is here that we can note the implications of his specific design of the Crakers. The aforementioned unwiring and rewiring of the Crakers' brains and the altered demands of their bodies in effect recreate the possible outcome of communication and society. Many of the forces that have led to the most horrifying events in human history are eliminated in the Crakers through the engineering of their drives, impulses, and needs.

It is important to note that the kind of society that would arise from this rewiring would be more likely to come about only if the Crakers are alone on Earth; they would have to completely replace humans 1.0. Were they to live in a vacuum, as it were, their design would ideally accomplish its purpose. In the novel, however, we see that many of the "safety valves" that Crake incorporates into the design of the Crakers are compromised because of the Crakers' contact with humans 1.0. For instance, in the time he spends with the Crakers Snow-

[5] There is considerable debate among rabbinic commentators and Kabbalists as to whether God created the universe *ex nihilo*. As this debate is outside the scope of this paper, for our purposes here suffice it to note that God's creation represents the materials of Crake's bio-engineering, and that God is attributed with creating life while Crake splices existing life.

man often has to explain why he does not look like they do and why he acts differently. He ascribes several natural phenomena to Oryx and Crake (such as lightning and thunder to the latter), and tells the Crakers stories – mythologies of origin. It is apparent that this contact with Snowman slowly "infects" the Crakers with needs and qualities of which they were designed to be free.

The Crakers' communication with Snowman leads them to form the kinds of ritual and worship that are associated with religion. In one instance, after Snowman's prolonged absence, the Crakers assemble an effigy of him. When Snowman returns the Crakers are overjoyed to see him, and tell him, "We made a picture of you, to help us send out our voices to you" (361). Snowman recalls Crake's concern: "*Watch out for art*, Crake used to say. *As soon as they start doing art, we're in trouble*. Symbolic thinking of any kind would signal downfall, in Crake's view. Next they'd be inventing idols, and funerals, and grave goods, and the afterlife, and sin … and kings, and then slavery and war" (361). In an ideal situation, it is possible that in the absence of contact with non-Crakers, the Crakers would remain free of the qualities, needs, and ideas of which Crake tries to keep them innocent. However, as it turns out, the Crakers are not left on the Earth in a vacuum, and so the ideal situation cannot be realized.

It is revealed toward the end of the novel that Crake's plan is indeed for the Crakers to replace *Homo sapiens* upon the latter's extinction due to the release of a pathogen. That is, while he and the rest of the engineers create humans 2.0 as prototypes so that potential parents can have children according to their design, Crake's real plan is to replace one species with the other. The eradication of humans 1.0 would accomplish the rewinding of history (without, of course, undoing the damage inflicted on the environment). The Crakers' unwiring and rewiring, on the other hand, is meant to ensure that history does not repeat itself. The unwiring and rewiring would prevent the repetition of history because it removes from humans the qualities that led to the ruin described in the novel. In fact, these are the same qualities than man is faced with upon his expulsion from Paradise. Not for naught is the Crakers' environment called Paradice: the Crakers are similar to man before his fall from Paradise. Both are naked and unashamed, both are pain-free and content, both need not labor so as to survive. Snowman, who first saw the Crakers when they were still in the Paradice Dome, and who lives close to them by the beach afterward, describes them:

> They seemed happy enough, or at least contented. They grazed, they slept, they sat for long hours doing what appeared to be nothing. The mothers nursed their babies, the young ones played. The men peed in a circle. One of the women came into her blue phase and the men performed their courtship dance. (339)

With no need for clothes, shelter, tools, or elaborate processes, the Crakers' lives resemble those of Adam and Eve in the Garden of Eden.

While the harmony and comfort experienced by Adam and Eve and the Crakers is similar, there is a difference in the level of participation required of them. The Crakers are asked of far less participation than man, even before his fall. Even though man's working of the earth was to be more grueling and painful after the fall (Rashi interprets "With the sweat of your face you shall eat bread (Gen. 3:19) as "After you toil with it very much" (Mid. Tadshei, Otzar Midrashim)), he was asked to till the soil even in Eden ("Now the Lord God took the man, and He placed him in the Garden of Eden to work it and to guard it" (Gen. 2:15)). Man's action in terms of prayer and development of vegetation was required. The Crakers, on the other hand, are not designed to collaborate or in any way participate in developing their surroundings. Moreover, the Crakers do not share Adam's role in characterizing and classifying the animals, that is, they are not portrayed as recognizing the intrinsic qualities of or having dominion over the animals that share their space. We see therefore that the Crakers participate in maintaining God's creation even less than Adam does, if at all. Without this attribute of naming and recognition, and without a role in the development of creation, the Crakers' role is not very different from that of the animals. There is yet another indication that the Crakers are designed to be more animal than human. In his commentary to Genesis 2:16, Munk notes,

> The fact that God first of all pronounces a prohibition instead of giving a command means implicitly that the one He addresses enjoys a new privilege: freedom [...] God no longer imposes a formal materialized order in the form of instinct as He did for the animals; He leaves man free to choose between obedience and dis-

> obedience and this, for the first time, introduces conscience and will as the means by which man will progress. Man thus becomes master of his own fate. (34)

The unwiring and rewiring of the Crakers' brains and the design of their bodies preclude not their freedom to choose, but the very need to make a choice. They are unknowingly bound by the "formal materialized order in the form of instinct," as are animals.

Yet a third reference to the biblical story supports the comparison of the Crakers to animals. Upon Adam and Eve's sin, God says, "Behold man has become like one of us, having the ability of knowing good and evil" (Gen. 3:22). Rashi interprets this verse as follows: "He is unique among the earthly beings, just as I am unique among the heavenly beings, and what is his uniqueness? To know good and evil, unlike the cattle and the beasts" (from Targum Jonathan, Gen. Rabbah 21:5). After they eat from the forbidden fruit, Adam and Eve become less like animals. The Crakers never gain "the ability of knowing good and evil" because their design does not allow them to access such a distinction. What Adam gains by eating from the fruit of the Tree of Knowledge of Good and Evil is something that the Crakers are designed to never attain: they are genetically engineered out of desire of such a fruit and what it represents, and will never have the knowledge it affords. The Crakers are thus incapable of wreaking the kind of havoc humans did upon the earth. Denied, through genetic engineering, the means by which to fall from Paradise, the Crakers are designed to remain there forever. In other words, populated by beings such as the Crakers, the Earth can remain Paradise.

In the setting of *Oryx and Crake*'s Paradice, the fruit of the forbidden tree – the means to fall – essentially has no equivalent. Accordingly, the narrative in this case is not one of prohibitions, commandments, temptations, and tests, as in Genesis 1 – 3. The wielding of biotechnology has rendered the novel's narrative of the Crakers one of sustainable cohabitation – sustainable harmony. In other words, it is a kind of inactive survival – a survival that does not depend on active participation on the part of the actors.

Munk writes that "The first sin destroyed the marvelous harmony of the beginning" (31). It is possible that the Crakers are designed to maintain this harmony specifically by *not* participating. The underlying notion is that if the harmony of creation is to be maintained, perhaps man, as we know him, as a participating creature, cannot exist.

He should be modified to be more like an animal, with less of an actively collaborative role in nature. Crake attempts to ensure a permanent state of Paradise – perpetual equilibrium and harmony – by removing any qualities that either lead to the fall, or to the actions of the already fallen. It would seem that a permanent state of this harmonious Paradise can only be attained if the Crakers are stripped of – more specifically – many of the qualities that would differentiate them as human. This raises the possibility that according to Crake, these eliminated qualities may be at odds with a state of equilibrium between man and his environment.

It appears that the being that participates less in his environment is also the being that exhibits few of the qualities and capabilities that are associated with the human. This is the type of being that Crake bets will survive in the next round of human life on Earth, and the being that will in turn allow for the Earth's survival. An analysis of man's relationship to nature in the biblical story foregrounds the purpose behind the Crakers' specific design, and suggests that this design is considered by Crake to be conducive to a perpetual harmonious state of pre-fall Paradise.

The etymology and semasiology of "paradise" and "Eden" reveal that they are characterized in terms of plant life. According to the entry for "Eden" in Oxford's *A Dictionary of the Bible*, "The garden of Eden came to be thought of as paradise, before the fall." The entry for "Paradise" in *The Oxford Companion to the Bible* states, "Originally a Persian word meaning 'park' or 'enclosure,' paradise first appears in the Septuagint with reference to the garden of Eden (Gen. 2) and became associated with a pristine state of perfection free of suffering." In the entry for "The Garden of Eden," the same reference states,

> A garden of trees and lush vegetation planted by God and occupied by Adam and Eve (Gen. 2–3). The meaning of the word "Eden" in Hebrew is uncertain. Some scholars connect it with a Sumerian word meaning "wilderness" or "plain," while others have proposed a derivation from the Hebrew word for "delight" or "pleasure." Thus, Eden came to be identified as an ideal garden of delight, or paradise.

We see that the distinctive, defining feature of paradise/Eden is plant life. That Crake seemingly chooses to return to and perpetuate paradise as the site of the euphonious existence of man and environment indicates that the relationship between humans and plant life is paramount in this harmony.

Works Cited

Atwood, Margaret. *Oryx and Crake*. New York: Anchor Books, 2004. Print.

—. *The Year of the Flood: A Novel*. New York: Nan A. Talese/Doubleday, 2009. Print.

Browning, W R. F. *A Dictionary of the Bible*. Oxford: Oxford University Press, 2010. Web. 3 July 2013.

"Chullin 60b". *Halakhah*. Tzvee Zahavy, 2010-2011. Web. 21 March 2013.

Metzger, Bruce M, and Michael D. Coogan. *The Oxford Companion to the Bible*. Oxford: Oxford University Press, 2004. Print.

Munk, Elie. *Call of the Torah: 1 Bereishis*. New York: Mesorah Publications, 2007. Print.

Scherman, Nosson. Overview. *The Wisdom in the Hebrew Alphabet: The Sacred Letters As a Guide to Jewish Deed and thought*. Michael L. Munk. New York: Mesorah Publications, 1983. 16-32. Print.

"The Complete Jewish Bible With Rashi Commentary." *Chabad*. Chabad. Web. 21 March 2013.

Iconic/Ironic Greenery: The Cultural Cultivation of Plants in Brecht Evens' *The Making Of*

Charlotte Pylyser

Abstract: This essay analyses the plant-strewn pages of Brecht Evens' graphic novel art parody *The Making Of* (2011) from an intertextual point of view that pays special attention to the Pattern and Decoration movement and to the element of kitsch. As a way of investigating the use of plant-life in art, this analysis shows how plants and flowers, even (and perhaps especially) when abundantly present in an artwork can be effaced in their symbolic and biological capacity and instead come to function in a purely instrumental manner. In light of the overwhelming silence of plants as plants in the graphic novel, this text raises epistemological questions pertaining to the knowability of plants to humans.

In her recent article "Art for Plant's Sake? Questioning Human Imperialism in the Age of Biotech" (2012), bioart scholar Monika Bakke describes how art can explore the post-natural condition that is typical of modern human-plant relationships and which is made explicit in the many biotechnological advancements, such as genetically modified crops, that contemporary man must learn to manage and to form an ethical opinion on. Bakke illustrates the creative power of biotechnology by describing Eduardo Kac's artwork *Edunia* (2003-2008), an instance of transgenesis, "the transfer of genetic material from one organism to another" (Bakke 10), which in this case involves the transfer of the artist's DNA onto a petunia plant whose pink flowers then symbolize the material and metaphysical "blooming" of a curious interspecies cross-breed. As Bakke goes on to explain, the creative gesture of the artist is at once also an invasive one as it changes the initial genetic make-up of the plant. This human-induced alteration does not only refer to the genetic engineering that takes place in modern day high-tech, highly sterilized, highly procedural laboratory setups, however. In the case of the petunia genus, it has precursors that date back to the nineteenth century when the wild petunia plant was imported from South America to be domesticated in botanical gardens and then cultivated into innumerable colourful variants by enthusiastic private garden owners (Bakke 11). Kac's fusing of himself with the plant then symbolizes an intensification of an age-old principle in the representation of the relationship between man and nature: man intervenes in (plant) life, (plant) life carries the mark of said intervention and perpetually confronts man with the agency and impact which the latter often seems to consider defining characteristics of his own kind.

Edunia walks a thin line between the performance and the critique of notions of "instrumentalisation, colonisation, separation and control" (Bakke 10) and while Bakke appears to consider this double state somewhat of a disappointment, it seems to me that it could be an interesting exercise to foreground this doubleness as a way of further exploring the conceptual and ethical challenges which plants pose to contemporary critical theory and practice. Kac's gesture also suggests a fundamental form of proximity between plants and humans, going so far as to reflect the mutual plant-human history which many proponents of ethical plant-human relations posit in a genetic opening up of those domains (in the Edunia, Kac and his petunias have a future together). Of course, if *Edunia* can be considered an instrumentalization of plant life, it can also be regarded as perpetuating the anthropocentric focus which is often characterized as accompanying said instrumentalization (indeed, Kac intervenes and leaves a symbolic and genetic mark on plants which can only return the favour on the symbolic plane – no mutual exchange of DNA took place). Again, it seems to me that this inconsistency can be fruitful ground for further investigation, particularly in terms of the ease with which the association between the ethical treatment of plants and their humanisation is formed. From Bakke's text it emerges quite clearly that the ethical conceptualization of plants is often considered as presupposing a movement that puts plants on the same level as humans in a sense that bestows upon plants almost all the functions which humans can perform. Plants can "talk," they can interact with the environment (Bakke mentions the quicksilver responses of the mimosa plant). "They 'experience the world in their own way' (Florianne Koechlin et al. quoted in Bakke 17), hence, they can sense, communicate and act. They are territorial beings, and this indicates some form of self-recognition, decision-making and communication. […] plants behave intelligently and […] some decision-making is taking place in the roots" (Bakke 17). The rationale behind allocating an experiential subjectivity proper to plants on the basis of traditionally humanized categorizations (in the case of territorialism one might even posit a connection with the instrumental, anthropocentric approach to human-plant ethics) certainly raises some questions. Indeed, if the ethical treatment of plants is contingent on their subjectivity and their subjectivity is dependent on their performance of traditionally human functions, one wonders if plants can truly be considered and treated in an ethical manner (as plants). It is not entirely clear whether the instability of the category of the human, the genetic component of which Bakke cleverly expands with an allusion to Bruno Latour's *We Have Never Been Modern* (1993) earlier on in her text[1], further problematizes or presents a solution to the humanization assumption underlying the discourse found in "Art for Plant's Sake? Questioning Human Imperialism in the Age of Biotech," but it certainly points towards the necessity of letting ethical concerns with regard to human-plant relationships enter in a dialogue with the epistemological questions revolving around the possibility (or impossibility) of the "knowability" of the nature of plants to man. Plant silence is then perhaps not such a straightforward matter as Bakke seems to suggest. Indeed, one could easily conceive of the inaccessibility of a realm of plant-to-plant interactions to man as an empowering phenomenon. In this light, Bakke's positing of man's "basic obligation of curiosity to plants" (Bakke 17) might perhaps be more productively read on a metalevel that opens questions far beyond the mystery of the lightning-speed movements of the mimosa plant and which loop all the way back to the conceptualization of man and the emancipatory movements which have marked and still mark human (or mutual human-plant) history.

In light of the question of green silence, it seems useful not only to pay attention to those artworks that attempt to criticize it by humanising plants, but also to consider art that unambiguously effaces greenery as greenery. Paying attention to non-green works of art and literature allows us not only to analyze the instrumentalization implicit in works such as Kac's on a larger (and more isolated) scale, it also offers us the opportunity to thematize a perspective that is absent from Bakke's article, that of the cultural cultivation of plant silence (as

[1] Bakke mentions that "'we have never been human.' After all, despite many metabolic differences, we share about eighteen percent of our genes with thale cress" (Bakke 10).

opposed to its biotechnological cultivation). Indeed, the question of how plant silence may be constructed in art and how it has been constructed throughout the course of art history is central to this text. Although a more general memetic[2] approach to this question could no doubt garner interesting insights, I propose to focus on the ideological and aesthetic practices used to instrumentalize plants in a particular case-study: the graphic novel[3] *The Making Of* (Drawn&Quarterly, 2012) by Flemish artist Brecht Evens. As this work is highly artistically intertextualized it will (indirectly) allow us to broaden our horizons towards other art works, traditions, and movements.

The Making Of was originally published in Flanders under its Dutch title *De Liefhebbers* (Oogachtend, 2010) and features a striking amount of plant renditions that in my reading of the book are characterised by an equally striking quality of functional transparency which I argue below is generated by three instances of instrumentalization on the level of plants: plants function as a site for the development of the intertextual art network and as a way for the artist to showcase his skill (image 1). The intertextually-charged plants function as a visual décor to the narrative in the style of both the miniature and the Pattern and Decoration movement of the late 1900s (image 2). Finally, these elements allows a parodic effect to be generated by the book that culminates in the presence of an uncanny garden gnome (image 3). Each function refigures plant life (which always seems to be considered as a collective entity in the book) into a specific shape that obscures the biological and the symbolic presence of the plants in question. Plants, in other words, do not function as plants. In the case of the art intertext, the plants that make up the references to paintings are reduced to landscapes and still lifes. The function of plants as décor/panorama sees them function merely as stylised patterns, and the tension between inside and outside as parodied in *The Making Of* on the level of

[2] The term "memetic" refers to the concept of the meme, which is to be understood here in the sense in which Richard Dawkins famously coined it in *The Selfish Gene* (1976): "[a meme is] the new replicator, [it is] a noun that conveys the idea of a unit of cultural transmission, or a unit of *imitation*" (Dawkins 192).

[3] A graphic novel is usually understood to be a narrative sequence of images and words in a one-shot book form, written by a complete author (an author that is responsible for both the drawings and the story of a given work) and oriented towards an adult audience. In the context of Evens' engagement with Art, however, it is important to add that the graphic novel is characterized by a culturally legitimatizing function.

art is reflected in the presence of the garden gnome, as this kitschy element of garden decoration evokes the cultivation (disciplining) of plants in the many bourgeois front yards dotted along the pages of the graphic novel in contrast with the panoramic woods through which the characters of the book move. Amplified by the importance of the author as a subject in the poetics of the graphic novel form (the authorial subject uses plants to generate an intertextual, a panoramic and a parodic effect that leaves no room for symbolic and biological plants and flowers in the book) it is truly possible to consider the investigation presented below as an inquiry into the connection between human-plant ethics and art at a remove of an explicit human-plant ethical context. Usually understood to be a narrative sequence of images and words in a one-shot book form, written by a complete author (an author that is responsible for both the drawings and the story of a given work) and oriented towards an adult audience, the graphic novel is considered from the point of view of its Flemish and highly institutionalized specificity in the investigation that follows. In order to familiarize the reader with *The Making Of*, I first provide a summary of its story before moving on to the role of art and of plants in the book.

Image 1: art intertext

Image 2: décor

Image 3: the construction of the garden gnome as well as the cover of *The Making Of*

***The Making Of*: Parody against Dichotomy**

In *The Making Of*, Brecht Evens tells us the clearly self-parodic tale of a city-dwelling artist who attends a biannual open-air art festival that is organized for the first time in Beerpoele, a small Flemish countryside town that is extremely excited to welcome its first professional artist. The artist in question, Pieterjan, functions as an everyman character who is fundamentally determined by two elements that are quite commonly associated with the figure of the artist and with the scenes in which he (women are mostly featured in supporting roles in *The Making Of*) moves: the unbearable lightness of being that characterizes the highly institutionalized art circuit in Belgium (and abroad, but as a phenomenon it is perhaps most typical of continental Europe) and the weight of the artist's grotesquely tormented ego. Thus, Pieterjan teaches art to children whose lack of talent both exasperates him and feeds a sense of comfortable superiority that allows him to deal with his students with a pitying patience that prevents him from ever really getting through to them. Similarly, he develops a relationship with a fashionably-bespectacled gallery owner whose mercantile instincts and enthusiastic response to a charismatic artistic rival secretly annoy him to the point of disdainful retreat. The relationship in question implodes as his grandly conceptual ode to small-minded nothingness at the biannual is struck by lightning. Indeed, the plot of *The Making Of* centers around the construction and destruction of a collective artistic project created *in situ*, as Pieterjan would say, that really amounts to a gigantic garden gnome made out of the kindergarten crafting material *par excellence*: *papier-mâché*.

Driven by the need to fulfill his role as an artist (to do his job), but resentful of the "missionary work" (Evens 2011[4]) that such an endeavour implies, Pieterjan directs an ensemble of colourful Beerpoele natives who are overjoyed to be able to construct scaffolding and tear up newspapers in the name of art (the garden gnome). All of the townfolk except one subject themselves ever so willingly to Pieterjan's vision of art. Only Dennis, the town's resident schizophrenic, who achieves peace of mind by relentlessly drawing spirals on every object he comes across, including the scaffolding for the grotesque garden gnome, disrupts Pieterjan's idyll of artistic professionalism. In a fit of ill-directed micro-managerial rage (scaffolding is functional with or

[4] No page numbers are indicated in *The Making Of*.

without spiral-decoration), Pieterjan insists that Dennis stop drawing curls on the wooden beams and exclaims that professionalism should not be sacrificed on the altar of inclusivity (Evens 2011). Dennis, significantly the only character who is described as expressing himself through art, is subsequently excluded from the project and locked in a make-shift cube where he spirals out of control up to the point of violence and later finds himself imprisoned in a medieval-style gaol that invites the reader to make connections to Evens' use of medieval iconographic and illumination techniques. Not only does the figure of the madman serve as a perfect foil to Pieterjan's absurd megalomania, he also foregrounds the question of the position of art with regard to the inside-outside dichotomy that characterises much of Western thought about the discipline and of the relevance of dichotomous thinking for answering that very question.

Pieterjan's outburst is of course ill-directed because the source of the frustration which he aims at Dennis lies in the general well-meaning naiveté and ignorance he encounters in his assignment to include the town into his artistic process. Pieterjan is frustrated not by the townfolk's lack of creativity (that lack mimics his own, as reflected by the preposterous gnome), but by their lack of understanding of the customs of the trade, their way of speaking about art, and their inability to act in accordance with what is expected of the art-savvy. The telephone conversations which the protagonist engages in with his gallery-keeping girlfriend, his life-line to the city, do reflect the *sérieux* which he misses in the countryside, but as often happens with customs and procedures, neither party appears to really enjoy these talks. Pieterjan finds some form of distraction in the arms of Cléo (after the muse presumably), a naively manipulative teenaged seductress with artistic ambitions whose unsuspecting use of difficult words turns him on. In his desire for her naive *sérieux*, his discourse of professionalism unravels and the utilitarian nature of his relationship with art is fully exposed. He uses his institutionalized artistic expertise to satisfy her need to feel special and accomplished and essentially talks her into bed. Cléo and Pieterjan's anticlimactic first and final night together reflects the wordplay inherent in the Dutch title of the book which relies on the double meaning of "liefhebbers" as a noun that signifies both "amateur" and "lover." Pieterjan's fizzled-out relationship with Cléo thus confirms his membership of exactly that group onto which he had bestowed his magnanimous condescension up until that time: the crafters, the wannabes, the garden variety of artists. If Dennis is Pieterjan's perfect foil in terms of the inside-outside dichotomy, then Cléo serves to demystify the exclusive relationship which art is often said to hold with the principle of autonomy.

It is safe to say that art, the creative process (which becomes a posture, a pose and an ideological production), and the positioning exercises of the field of art constitute the central thematic cluster in the story of *The Making Of*. But art is important to the book in other ways as well, including the style in which the book is painted (the artwork in the graphic novel is primarily rendered through watercolors) and the intertextual/interartistic art references that are strewn across its pages. It is in these elements that the connection between art and plant-life begins to unfold.

The Suggestion of Art: Instrument, Discourse, Ideology

In Flanders and abroad, Brecht Evens is lauded for his trademark watercolor style that can best be described as lush, luminescent, and colourful. Evens has a tendency to play with perspective (especially when it comes to buildings and structures) and with the opacity of his ecoline-paintings. He often includes so-called splash pages, pages that are taken up almost entirely by a single image, or spreads, two consecutive pages that contain a single image, into his graphic novels so as to regulate the rhythm of the narrative, to foreground certain elements, or to indicate that the narrative is to be read on a different level. One such a level is that of the dream. Dreams are rendered in a black and white style the simplicity of which contrasts very vividly with the exuberance of Evens' watercolors. Another level is that of the art intertext. Even's use of splash pages, especially in the case of his appropriation of existing artworks such as Hokusai's iconic woodblock print *The Great Wave off Kanagawa* (1830-1833), Matisse's fauvist *Landscape at Collioure* (1905), or the late medieval Flemish unicorn tapestries, always carries the connotation of the picture frame. Even when Evens' mise-en-page follows a more traditional comics format, he never opts for a regular grid pattern with closed-frame panels, rendering his edgeless panels open to a painterly or artistic interpretation as well. The author confirms that potentiality by evoking paintings such as *Le bonheur de vivre* (1905-1906), again by Matisse, and various cubistic still lifes by Braque in separate panels, but he also expands the iconic references to paintings by repeating stylistic characteristics of the artists and artworks that we have mentioned in panels that depict scenes that are less obviously connected to the artworks in question. One striking instance of this approach can be observed in Evens' transferral of the color blotch constellation which we recognise in the landscape of *Le bonheur de vivre* onto a different open-air scene that then interestingly becomes instantly recognizable as somehow connected to the art paradigm. Clear echoes of Seurat's pointillism and hints of Henri Rousseau's exotic landscapes add to the intermedial network, but it is possible to detect traces of other 19th and early 20th century artists as well, such as Cézanne and Van Gogh. Japanese woodblock printing and medieval iconography similarly form a graphic reference throughout the book. Finally, Evens' merry band of amateurs contains certain art-related figures in the form of an insecure balloon-animal-creating Pierrot (the sad clown which originated in the Commedia dell'Arte and has come to function as a double of the artist) and a perpetually heart-broken action figure version of the protagonist of Munch's *The Scream* series (1893-1910) called Valentine. The effect of this network of art references is that the suggestion of art is present at all times in *The Making Of.*

If we take a closer look at the nature of the suggestion which Evens constructs in his book, we see that the intertextual references mimic and support the role of art as we have observed it in our character-driven analysis. In *The Making Of,* art is instrument, discourse, and ideology. Evens does not interact with the artists he references on their terms, neither when it comes to what it is that his references are exploring or expressing (in this context the lack of attention which his highly postmodern intertextual, parodic and metareflective graphic novel bestows upon postmodern art is doubly striking, although this is quite a typical aspect of the postmodern method), nor in terms of the graphic reflection of that expression and exploration. Matisse and company are recycled, assembled, stuck together, and applied strategically in places where Evens wants to convey some minimally recognizable trace of the art intertext and serve as a way of keeping the many outdoor landscapes, which prominently feature all sorts of plants, in the book interesting through a mix of their visual ingenuity, attractiveness, and variety and through the intertextual referential frame they evoke. In opting for an instrumental use of art within a graphic novel art parody (note that the graphic novel is strongly tied to the comics medium which in turn may be linked with certain mass productive practices typical of the culture industry of the twentieth century) – in capitalizing on some of the properties of the intertext, in other words – Evens may be said to play with the critique of notorious, but not quite contemporary, mass cultural production critics such as Adorno who complained in his "Culture Industry Reconsidered" (1975) that "the cultural commodities of the industry are governed, as Brecht and Suhrkamp expressed it thirty years ago, by the principle of their realization as value, and not by their own specific content and harmonious formation" (Adorno 13).

The Making Of Plants: Panoramic Propulsion

I have spent some time sketching the art problematic in *The Making Of* exactly because it is primarily in the graphic rendition of plants and flowers that Evens manages to cultivate its complexity. Evens' plants evoke 1) the difficulty of maintaining outside/inside and autonomy/heteronomy dichotomies with regard to (the) contemporary art (world) in terms of the specificity of the graphic novel in relation to art, literature and comics 2) the parodic effect of the narrative (which relies heavily on the cultivation of kitsch in the use of plants and gardens) and 3) the nature of Evens' use of the art intertext, which constantly oscillates between postmodern meta-self-reflexive cleverness and a stubborn unwillingness to commit to any sort of grander narrative. In reverse order, we could posit that it is Evens' use of the art intertext that helps create the parodic effect which in turn serves to question the dichotomies which we have introduced. It is in this order that I propose to have a look at the way in which Evens utilizes greenery in *The Making Of*.

Merely flipping through the pages of *The Making Of* will confirm the overwhelming presence of greenery in the book. The silhouette of plants, trees and flowers and especially landscapes or still lifes filled with plants, trees and flowers (framed plants and flowers) visually dominates the graphic novel. Our eyes wander through leafy, dark forests, over swampy fern-filled grounds, through meticulously crafted front yards, and onto bridges that cross speech balloon ponds in fairy tale parks. We pass by warmly-colored sunset shores, disappear into tangles of multi-coloured underwater algae, and are tempted by the dark-blue night to walk a walk of shame along the greenhouse of Cléo's eccentric orchid-cultivating uncle – a nod to *Twin Peak's* Harold Smith perhaps – and into the arms of Pieterjan's uncanny sculptural garden gnome.

It should not come as a surprise that my description of plant-life in the book reads like something of a panorama considering Evens' ambition to sketch an environment in his graphic novels (and his background in illustration arguably). But one can also sense a strong element of sequential propulsion in the image that I have conveyed in the previous paragraph. Indeed, on various occasions we see the characters in Evens' tragicomedy travel linearly through the lushly decorated pages which he has composed. Sometimes this happens indirectly, when the liminality of a scene is repeated on an adjoining page and features the same figure, for instance. This technique is evident in Evens' introduction of an apparently innocuous cat who is shown moving from the woods into civilization on a splash page and then charms its way into the garden shed where the protagonist is put up for the duration of the festival on the next page. In contrast with the splash page, the latter page is divided into borderless panels, making the transition of the cat from the forest and into the ordered universe of the carefully cultivated garden a challenge to the reader's ability to follow the sequential integration of the story. Such devices create an interesting and quite pervasive undercurrent of readerly disorientation.

Other times, the characters' movement literally starts and stops at opposite edges of a page or spread, and, in some cases, topples over onto the next page or panel. Thus we find that Pieterjan and Chloé walk through a garden of Eden setting on an almost cinematic trajectory that first depicts the couple in the upper left-hand corner of the exotic spread and then each time foregrounds a moment of their trajectory until the final still brings our eye to the bottom right corner of the second page of the spread. The suggested propulsion is emphasised by the connection of the speech balloons that contain their dialogue, but in contrast with a true cinematic effect, each moment on the path which materializes by grace of the expression of certain points on the trajectory also coincides with a visually explicated position which the point in question takes up in the context of the spread's tabular workings. The above technique is one that can be quite strongly linked to the author ever since he made use of it on the cover of *The Wrong Place* (he did so on two separate occasions, both the cover illustration of the original Dutch version and of the translated versions make use of the idea of propulsion). As Greice Schneider shows in her analysis of *The Wrong Place*, this form of spatial propulsion entails an unambiguous element of temporal organization (the trajectory of the characters on the page implies a temporal order, sometimes also a causal one) that supplies much-needed reading direction to pages that otherwise challenge a great deal of the panel-driven sequentiality typical of comics (Schneider). In *The Making Of*, such instances are somewhat more rare than in *The Wrong Place*, and the intensification of the transitions from the splash pages and spreads that foreground the tabular workings of the comics pages to those pages that follow a more classic comics lay-out pattern, as well as the tension that results from the difficulty of successfully establishing such a transition, greatly foregrounds the role of the sequence in the narrative project of the former book. While the deployment of sequentiality as a connecting and ordering mechanism contains some glitches in the book, the intention of the representation of visual progress as well as its narrative reflection in terms of the linear structure of the *Bildungs* plot that characterises *The Making Of* is massively present and frustrates any attempt at book-length free-roaming which our eye might undertake – the panorama can never truly be viewed at a single glance.

Thus the effect of Evens' panoramic propulsion is a double one. On the one hand, the flowery pages of *The Making Of* gain a semblance of autonomy in the sense that they suggest that the setting through which the characters move constitutes an environment (as distinguished from a story world) unto itself. On the other hand, this autonomy is always restricted in its potential by the needs of the sequential narrative and the requirements for intelligible representation of sequential order. The challenges which the narrative poses to the panoramic whole are most obviously foregrounded in the constant style register ruptures that are necessary building blocks for Evens' art intertext, which is crucial to the parodic effect which the graphic novel generates. We move from meticulously cut-out miniature-esque flower fields over broad, bright fauvist streaks of grass to pointillistic beaches. In deploying his intertextual network thusly, Evens may be said to opt for a method that foregrounds greenery in order to have it work primarily as background or décor, rather than as an environment.

Evens' use of plants as décor invites connections to the Pattern and Decoration (P&D) movement which Gregory Fuller describes as follows in his book on kitsch art: "The P&D artists attempted to undo the division between High Art and décor. Traditional fabrics and materials were used often and folk art, ethnic décor and traditional crafts were rehabilitated as simultaneously old and new 'human' motifs" (Fuller 14;16, my translation). In contrast to Evens' use of the art intertext which we have introduced above, the author does not seem to refer to specific P&D artworks, but rather makes use of the general aesthetic of the movement. In doing so, especially in his elaboration of the floral motives which grace the pages of *The Making Of*, Evens echoes the central P&D concern with the division between High art (specifically the institutionalized art world in his case) and amateur crafts, between the in/out and autonomy/ heteronomy dichotomies pertaining to art. Indeed, if plants and flowers are involved in the development of the art intertext which we have introduced above, greenery also becomes the site for the evocation of a kitschy (in the broad sense of the term) amateur aesthetic that refers to the small-town hobbyists which the protagonist encounters in Beerpoele. At times the same patch of flowers or plants may even function as the venue for both, as is the case for Evens' appropriation of one of the unicorn tapestries mentioned above (image 1).

While Evens refers specifically to that individual hanging of the medieval unicorn narrative that shows the elusive unicorn corralled and which is aptly called "The Unicorn in Captivity,"[5] it is the intricate miniature floral technique which functions as a background to the entire series that is repeated several times throughout the graphic novel. Without the medieval context or the immediate intertextual reference, however, the symbolic connotation of the medieval hand-woven flowers, which often refer to fertility in the original hanging, is lost. Evens' stylized miniature flowers, which are rendered in a near-translucent negative watercolor technique, are also much harder to taxonomize and in combination with the placement of the flowers on the page as decorative flourishes or edging (frames), the floral presence in *The Making Of* is most reminiscent of the delicate flowers typical of a certain style of stationary paper design. In addition to a décor (as a panoramic narrative series the Unicorn Tapestries may be considered an inspiration on this level too), flowers thus also function as decoration in Evens' graphic novel, which arguably adds to their

[5] The Unicorn Tapestries are reproduced and explained further on the website of the Metropolitan Museum which houses the originals in their collection: http://www.metmuseum.org/collections/search-the-collections/70007568.

biological and symbolic transparency. Indeed, Evens' flowers may be said to produce the suggestion of kitsch by analogy with the suggestion of art which we have commented on before.

Evens' parodic commentary unfolds beautifully in his use of "The Unicorn in Captivity," which is after all not only a top piece in one of the most renowned museums on earth, but also an anonymous and painstakingly handcrafted artefact (a wall-hanging of a different sort than a painting) made on commission (the common mode in which art was produced up until the 18th century). The instrumental stationary effect which he achieves by isolating and reproducing a decorative pattern belonging to the otherwise highly iconic and symbolic artwork is particularly interesting when we consider that the fabled and stately unicorn, which in *The Making Of* has been transformed into a peculiar crossbreed of a work horse and a fairground attraction, shares its function as the object of a great quest with the preposterous garden gnome (image 3) worshipped by Pieterjan and his disciples. In Evens' sardonic appropriation of the instrumentalization of plants in the décor and decoration traditions, the practice of the cultural cultivation of plants (its richness as well as its limitations) is made eminently visible.

Works Cited

Adorno, Theodor W. "Culture Industry Reconsidered" *New German Critique* 6 (1975): 12-19. Print.

Bakke, Monika. "Art for Plant's Sake? Questioning Human Imperialism in the Age of Biotech." *Parallax* 18.4 (2012): 9-25. Print.

Conard, Sébastien. "Alles voor het oog? Over *De Liefhebbers* van Brecht Evens." *Rekto:Verso: Tijdschrift voor cultuur en kritiek.* 50 (2012): 14-17. Print.

Evens, Brecht. *De liefhebbers* [*The Making Of*]. Leuven: Oogachtend, 2011. Print.

Schneider, Greice. "The Wrong Place – Brecht Evens." *The Comics Grid.* 7 Feb. 2011. Web. 15 Mar. 2013. Print.

A Return to Transcendentalism in the Twentieth Century: Emerging Plant-Sympathy in *The Little Shop of Horrors*

Stephanie Lim

Abstract: Within the last several decades, Hollywood has experienced a surge of apocalyptic, man-versus-plant, or "killer-plant," narratives, with films like *The Thing* (1951), *Invasion of the Body Snatchers* (1956), and, more recently, *The Ruins* (2008) hitting big screens. When studied more closely, these narratives seemingly reveal a growing fear that human actions towards Earth's greenery – cutting down forests and using chemicals on plants, for instance – are much more destructive than we once thought. However, while most films simply produce antagonistic attitudes towards plant, *The Little Shop of Horrors* (1960) showcases a more symbiotic relationship between man and plant.

Ralph Waldo Emerson once proposed the existence of "an occult relationship between man and vegetable" (494), and this mysterious intersection between humankind and nature – plants, specifically – has been explored in many American texts, ranging from Nathaniel Hawthorne's 1844 short story "Rappaccini's Daughter" to Hollywood movies like the 1960 comedy *The Little Shop of Horrors*. Even though Emerson's nineteenth-century Transcendentalism essentially provides an optimistic and amiable view towards nature, however, a consistent message of fear and anxiety about vegetation has been produced in American audiences for over a century, as Hollywood has been obsessed with "killer plant narratives" in which humans find themselves in gruesome, life-and-death struggles with Earth's greenery. But what if a seemingly vicious, man-eating plant was simply a misunderstood organism, actually providing economic stability and social growth to an entire community? What would such a plant suggest, as Western societies continue to advance industrially and technologically? Although Hollywood's fascination with the killer plant narrative has already stretched into the twenty-first century, a critical study of Seymour Krelboyne's enigmatic relationship to the half-butterwort, half-Venus Flytrap Audrey Junior in *The Little Shop of Horrors* reveals an emerging "plant-sympathy," or compassionate attitude towards plants, in America during the 1960s.

Above all, *The Little Shop of Horrors* signals a distinct turning point in American socio-political ideology regarding humankind's relationship to Mother Earth. Understanding the sense in which *The Little Shop of Horrors* is not merely another Hollywood movie utilizing the infamous killer plant narrative but, instead, is a representation of American fears and anticipations of larger socio-economic issues to come requires an exploration of the larger social atmosphere during the mid-1900s. It would at least appear to be more than sheer coincidence that the American environmental movement, roots of which are found in Transcendentalism and the Progressive Era, also gained its momentum in the 1960s. When considered in light of the other social developments happening simultaneously in America in the 1960s, *The Little Shop of Horrors* certainly becomes exceedingly symbolic of the decade's ideology, and the film quite possibly represents a turning point in society towards a more positive and sympathetic attitude towards plant life.

This essay is demarcated by three sections: a history of killer plant narratives in American literature and media, an understanding of the unique positioning of *The Little Shop of Horrors*, and a brief discussion regarding the nation's Green Movement, which coincides, in time and in ideology, with the 1960 film. Perhaps it is time to end the epoch of unease and terror associated with nature and, instead, shine a more progressive light on killer plant narratives found in American cultural texts.

From Myth to Mainstream: A History of Infamous Killer Plant Narratives

> The atrocious cannibal tree that had been so inert and dead came to sudden savage life. The slender delicate palpi, with the fury of starved serpents, quivered a moment over [the woman's] head, then as if instinct with demoniac intelligence fastened upon her in sudden coils round and round her neck and arms; then while her awful screams and yet more awful laughter rose wildly to be instantly strangled down again into a gurgling moan, the tendrils one after another, like green serpents, with brutal energy and infernal rapidity, rose, retracted themselves, and

> wrapped her about in fold after fold, ever tightening with cruel swiftness and the savage tenacity of anacondas fastening upon their prey. It was the barbarity of the Laocoön without its beauty - this strange horrible murder. (Osborn 8)

Chase Salmon Osborn, the governor of Michigan in the early 1900s, published the above depiction of a man-eating plant in his 1924 book, *Madagascar, Land of the Man-Eating Tree*, in which he tells of his travels across the island country of Madagascar and includes detailed written descriptions of the island and its inhabitants, as well as a number of illustrations and maps of the island. As Osborn explains, the account quoted above was originally provided in 1880 by German explorer Carle Liche, who first wrote about the man-eating tree in the New York *World* and then again in the *South Australian Register.* While the book offers readers a detailed historical, political, and geographical overview of the island, in addition to information regarding the Malagasy culture, Osborn was never able to find the so-called man-eating tree; in fact, he admits, "in travelling from one end of Madagascar to the other a thousand miles and across the great island, many times traversing the nearly four hundred miles of breadth, I did not see a man-eating tree" (2). This, however, is one of the first documented accounts of a man-eating plant in the Western world.

Osborn, of course, was neither the first nor the last person to explore the land of Madagascar in search of Liche's incredible discovery, nor was he the only one to ask the Malagasy people about the tree. Between 1850 and 1950, explorers like Frank Vincent, Ralph Linton, and Willy Ley continued to encounter believers of the mythological organism and attempted to conduct their own research in search of the truth. Ley eventually concluded in 1955:

> Of course the man-eating tree does not exist. There is no such tribe. The actual natives of Madagascar do not have such a legend. But at one time somebody made up the hoax, which was put into the only existing local magazine, possibly as a joke of some kind for the amusement of the readers who knew better. But it then got out of hand and the perpetrators thought it best to keep quiet. And if Mr. Chase Salmon Osborn, browsing around on the spot, had not resurrected it, the whole thing would have been forgotten a long time ago. (182)

The impact of the legendary hoax, however, and the overall intrigue surrounding violent plants, has continued to live on through various cultural mediums. To this day, killer plant narratives have made a big dent in popular culture, with aggressive shrubbery landing lead roles in movies, television shows, books, and video games. The following section addresses major killer plant narratives found within twentieth- and twenty-first century mass media, in addition to the inherent ideology presented through the texts.

Killer Plant Narratives in Literature. America's obsession with violent, nature-versus-man stories has taken on various forms in fictional novels and short stories. Some plants are not physically aggressive like much of today's killer plants are, but instead are portrayed as passive in character. Nathaniel Hawthorne's "Rappaccini's Daughter," for example, takes a subtle approach towards killer plants, depicting a young woman, Beatrice, who becomes resistant to her father's experimental, poisonous plants and who has become poisonous herself. The protagonist, Giovanni, having fallen in love with Beatrice, eventually becomes poisonous as well, with the ability to breathe poison onto insects and instantly kill them. John Collier, whose short story, "Green Thoughts" (1947), is thought to be the main influence for *The Little Shop of Horrors*, also paints the story of a passive plant who consumes the protagonist, his cousin, and his cousin's cat. The protagonist, however, does not die and instead goes into what the narrator describes as a "vegetable lethargy" (79). Likewise, Robert Ayre's "Mr. Sycamore" (1947) is the story of a man who longingly aspires to become a tree, literally, standing out in his yard for nights on end, despite his wife's attempts to reason with him.

Naturally, though, there is literature depicting fictional plants that are extremely aggressive in nature. *Beyond the Deepwoods* (1998), Paul Stewart and Chris Riddell's fantasy novel, takes the opposite approach, depicting an explicitly violent plant, Bloodoak, that attempts to devour the novel's hero, Twig. Author Annie Proulx follows suit in her short story "The Sagebrush Kid" (2008), in which a couple tends to their sagebrush plant much like they would a child, caring for it and feeding it human food; but, like a child, the sagebrush grows up and eventually begins desiring more substantial meals, mainly in the form of adult human beings.

Killer Plant Narratives in Modern Mainstream Media. Killer plants have made the biggest impact on contemporary mainstream media, like popular television, films, and even video games. The world of electronic games in 2009 saw the release of PopCap Games' *Plants vs. Zombies*, in which players use various types of plants to fend off zombie attacks on a house. Game play is both defensive and offensive in nature, with players being able to use plants to protect the house or to gain energy from the sun, as well as to aggressively attack the zombies.

In 2008, *Den of Geek*, a famous "geek" culture website originally based in the UK that reviews media such as movies, books, television shows, comics, and video games, compiled a list of the top ten killer plants which have made an impact on the Hollywood big screen. Many of these films have been remade twice over, and a majority of the films have been digitally transferred from their original VHS formats to DVD formats, all of which is indicative of the resilience killer plant narratives have had into the twenty-first century. In what follows, I will briefly examine each of *Den of Geek's* highlighted killer plants, in chronological order[1], before synthesizing the ideological implications these killer plant narratives express:

- *The Thing From Another World* (1951 and 1982, science fiction), also known as simply *The Thing*, presents an alien being with plant-like characteristics which has been uncovered and awakened after a period of time.
- *Invasion Of The Body Snatchers* (1956, 1978, and 2007, science fiction) follows in the same vein as *The Thing*, depicting xenobotanical, or alien plant, beings that take over the human bodies they embed themselves in.
- *Little Shop Of Horrors* (1960 and 1986, comedy film and musical) features a young man who crossbreeds two plants, creating a man-eating, venus-flytrap-like plant which requires human blood and, eventually, human bodies to grow and survive.
- *Day Of The Triffids* (1962 and 1981 TV, apocalyptic science fiction) introduces triffids, fictional plants first created by John Wyndham in his 1951 novel of the same name; triffids function

[1] For purposes of this essay, I have chosen to go through the movies in chronological order from the earliest to the most recent and have also included, in parentheses, the genre(s) in which each movie is normally categorized. However, it should be noted that *Den of Geek* presents their list of the top ten in a regular "countdown" order based on their own criteria, which is explained in detail online.

independently, are self-reliant and self-sustaining, poisonous, and prolific. The triffids pose a significant threat to humans because of their presumed capacity to take over the world.

- *Dr. Terror's House Of Horrors* (1965, horror anthology) features a segment entitled "Creeping Vine," one of five horror stories. In it, a plant responds aggressively whenever humans attempt to cut it down.
- In the serial *The Seeds Of Doom* (1976 TV, gothic horror/science fiction), part of the original, British *Doctor Who* series, a vegetation pod stings one of the main characters, Winlett, who then becomes engulfed in green fungus. The pods ultimately get into his bloodstream; he no longer has human blood platelets but prokaryotes (formerly Schizomycetes), or plant bacteria. The serial also shows the transformation of Winlett into a Krynoid, an alien plant form that destroys animal life, and the humans are eventually seen turning on each other in order to survive.
- *Attack Of The Killer Tomatoes*[2] (1978, comedy horror) humorously depicts aggressive vegetation in the form of killer vegetables (tomatoes), which corner humans and eat them alive. The film also shows American military efforts to thwart, and eventually destroy, the aggressive tomatoes; however, at the end of the film, carrots are shown as the next vegetable group set to take over the world.
- *Creepshow* (1982, horror anthology) features the short story, "The Lonesome Death of Jordy Verrill," in which a substance from an alien meteorite causes a plant-like organism to begin growing on body of Jordy. The film's ending implies that the vegetation will continue to spread on both mankind and the earth.
- *The Guardian* (1990, horror film) commences with the following historical context: "For thousands of years a religious order known as the druids worshipped trees, sometimes even sacrificing human beings to them. To these worshippers, every tree has its guardian spirit. Most are aligned with goodness and life, but some embody powers of darkness and evil." In the film, a half-tree, half-human character is seen sacrificing human life to a sacred tree, both in order to help sustain the tree's life and as an expression of devotion and reverence to the tree.

---

[2] *Dr. Terror's House Of Horrors* and *Attack Of The Killer Tomatoes*, while not in the top ten, are named by *Den of Geek* as honorable mentions.

- In *Minority Report* (2002, neo-noir/science fiction), Tom Cruise's character is scratched by baneberry, a plant that is said to be poisonous and will kill him within seconds.
- *The Happening* (2008, supernatural thriller) illustrates aggressive tactics utilized by plants. Because plants feel threatened by mankind, who attempt to cut plants down, the plants respond by releasing toxins into the atmosphere causing humans to commit suicide.
- *The Ruins* (2008, horror film) shows predatory vines which aggressively attack and take over human life, ultimately forcing humans to turn on each other.

The Ideological Implications of Man-Eating Organisms. Keeping the aforementioned films in mind, three typical characteristics of the killer plants in killer plant narratives are worth mentioning. First, most of the plants are depicted as alive and aware, with many functioning at high, human-like intellectual levels. Some also have great physical abilities and can function independently, although the plants always require some sort of nourishment, normally in the form of humans, in order to thrive and for sustenance. Second, several of the films strongly emphasize that the aggression shown by plant life is a defensive response. That is to say, physical threats by humans towards a plant's existence, such as cutting down a tree or trimming the leaves of a plant, are shown as the primary trigger of a plant's violent tendencies towards humankind. The plants, if not alien beings, did not initially begin killing humans to take over the world but, instead, are retaliating against human behavior. Third, killer plants are almost always portrayed as xenobotanical in nature, a very sure-fire rhetorical strategy of movie makers. In fact, the way in which plants are portrayed in these films is quite telling of the socio-political ideology over the last few decades, mainly the fear of Mother Nature's vengeance against humans. Marc Jensen highlights the function of the monster movie genre and specifically what the genre articulates about social attitudes, contending that, "[s]ince its invention, the monster movie genre has been used as a vehicle to articulate deeply rooted societal fears. Embodying the subject of fear as a physical manifestation on the screen, the action of the movie serves as metaphor or allegory. Monster movies often invoke a xenophobic fear of the "other" (51). We, the audience, will naturally have a sense of hostility towards the unknown, or the mysterious, especially when we believe that our lives or our own world is being threatened. By presenting a plant as coming from an unknown, alien world, audiences automatically "other" the plant beings, and, thus, the killer plant fear is constructed.

Additionally, there is a recurring theme of killer plants literally taking over human bodies. The takeover does not happen purely physically – that is, death – but can also happen mentally, such as charming humans or distorting the human mind in order to turn humans against each other. This closely mirrors zombie apocalypse movies, wherein the zombie "disease" forces humans to turn against each other and causes an overall collapse of society. The hostile takeover theme showcases an unconscious fear of plants controlling human life and the human mind, the overpowering of human beings, and, ultimately, the crumbling of civilization.

While it is clear that killer plant narratives almost never depict a friendly relationship between man and plant, the endings of the films are also important to consider. With what type of ending do the movies leave audience members and with what intention or in what context are the movies themselves framed? (Almost all of the movies I have thus far named are dramatic horror films, and well-known filmmakers and film-writers associated with the horror genre are associated with the films, such as M. Night Shyamalan, Stephen King, and Steven Spielberg.) What most killer plant narratives actually teach us is that man-plant wars are never won by showing more aggression to the plants. Killer plant film endings normally take one of two routes: the ambiguous "to be continued" route, wherein man defeats the plant, but the plant, at the very end, is shown to have resurrected somehow and will continue to live on, or the clearly apocalyptic route, wherein the plant explicitly defeats man, with no hope of man's survival. Regardless of the route a movie takes, it is rare to find a film that shows man as having completely thwarted a hostile plant-takeover. In many cases, in fact, man is not the last life form standing, entirely overwhelmed by nature's aggressive attacks. Therefore, man is never "safe" from nature, a condition that perpetuates human hostility towards plant life.

Appealing to the Masses: Why Killer Plant Narratives Survive. The majority of killer plant narratives present to audiences a strong, hostile tension in the relationship between man and nature, essentially illustrating the ideology of the twentieth and twenty-first centuries in America: man is a dominant being, whereas nature is to be controlled. But why use a killer plant? Science historian Stephen Jay Gould suggests:

> Much of our fascination with 'strange cases' lies in their abrogation of accepted dualisms based on dominance [...] As an obvious example, and paragon of this literature, carnivorous plants have always elicited primal intrigue – and the bigger and more taxonomically 'advanced' the prey, the more we feel the weirdness. We yawn when a Venus's flytrap ensnares a mosquito, but shiver with substantial discomfort when a large pitcher plant devours a bird or rodent. (394-5)

Movie makers use an over-dramatized notion of killer plants which, in reality, would never harm humans; in fact, not only are plants actually beneficial to humans, but they are also in a cooperative relationship with human life. Take, as the main example, the importance of trees in the removal of carbon dioxide from the atmosphere and, as a result, the production of oxygen for humans. Lipske and Noona also emphasize the lack of antagonistic struggle between humans and plants with an anecdote: "'Did you ever see that Bugs Bunny cartoon where he cuts off Florida at the top of the peninsula and pushes it out to sea?' he asks. 'And it floats away?' [Late Auburn University professor George Folkerts] smiles, knowing full well that only in the movies do killer plants ever get the upper hand." Conceivably, however scary a killer plant may look on screen, audience members always return home knowing that the flowers in their backyards cannot possibly launch a war against them.

The reason killer plants seem so attractive to movie makers and movie viewers alike is the manner in which they play on the socio-ecological hierarchy. Gould brands this hierarchal disruption as the "reversal of established orders," an intriguing understanding of how narratives, such as killer plant stories, function in Western society and, more importantly, why they persist. Gould calls into question, in particular, the practice of taxonomy, asserting that, "[i]n creating dualisms, we divide a subject into two contrasting categories; in imposing hierarchy upon these dualisms, we judge one category as superior, the other as inferior ... We often divide the world ecologically into predators and prey, or anatomically into complicated and dominant 'higher' animals versus simpler and subservient "lower' forms" (394). In the case of the man-nature binary relationship, man is always superior to nature. This hierarchy is, of course, challenged by killer plant narratives and, thus, makes for very intriguing, and sometimes horrifying, storylines. Killer plant narratives, however, do not only attempt to reverse established orders present in the real world but also signal significant changes in thought, and it is these ideological implications conveyed by the storylines that we must seek to investigate.

Love at First Sight: A Symbiotic Relationship between Man and Plant

> We have come to the edge of a world of which we have no experience, and where all our preconceptions must be recast (Thompson 77).

While a majority of the books, films, and other media mentioned offers audiences an antagonistic attitude towards greenery, four storylines in particular stand out as offering more than just tales of terror and apocalypse: "Rappaccini's Daughter," "Green Thoughts," "Mr. Sycamore," and *Little Shop of Horrors*. Unlike the bulk of their counterparts, these stories approach plant life with a very distinctive attitude towards plants, one that shows the relationship between man and plant as much more complex than the average killer-plant-as-evil narrative, forcing audiences to reconsider their antipathy towards man-eating plants. Complexities arise, mainly because the attitude and emotions conveyed in each tale are neither completely positive nor completely negative, perhaps due to the fact that these stories follow more closely the mind of the protagonist who is consumed, as opposed to the bystanders who helplessly watch as another human gets eaten. Moreover, these stories focus more on human protagonists than the actions of the carnivorous plants. Thus, audiences are forced to think beyond the binaries of "good" and "evil"; our preconceptions of the man-on-top socio-ecological hierarchy are put to a halt, and we are even de-stigmatized to death and murder.

"Rappaccini's Daughter" offers two particularly intriguing views of human-plant relationships. In several instances throughout the story, Beatrice herself explicitly refers to a particular shrub in her father's garden as her "sister," even "[opening] her arms as if to embrace [the plant]," which illustrates the familial bond Beatrice feels towards the shrub. Giovanni, too, while observing Beatrice from afar, "[doubts] whether [she] were a girl tending her favorite flower, or one sister performing the duties of affection to another." Giovanni goes even further, perceiving an inherent likeness between human and plant: "[f]lower and maiden were different, and yet the same, and fraught with some strange peril in either shape." Hawthorne's tale, then, elucidates the notions that a human can have an intimate, platonic, relationship with a plant and that a human can have a physical likeness with a plant.

With a similar attitude towards plants, "Green Thoughts" also offers a non-hostile, though still unnatural and complicated, entanglement between man and plant. The orchid that, unfortunately, devours Cousin Jane and her cat, also ends up devouring Mr. Mannering. It is important to note, however, that the "murders" of both Mannering's cousin and the cousin's cat are not witnessed by the narrator, and Mannering, while also eaten by the orchid, is too much in a state of shock to put up a fight. These otherwise silent meals render the plant as merely hungry as opposed to entirely violent and aggressive. Furthermore, although Mannering is devoured, he is not necessarily angry about the situation. By the end of the story, Mannering's mentality has matched his physical form. The narrator explains, "because he *was* now a vegetable, he responded with a vegetable reaction" (77); he becomes vegetable in mind and spirit. When his nephew disrespects the elegant furniture in his house, Mannering's emotional response is "neither fury nor mortification"; "his ego was not reduced but concentrated, his serene, flowerlike indifference towards the ill-usage of his furniture was balanced by the absorbed, flowerlike single-mindedness of his terror at the thought of similar ill-usage directed towards himself" (87). Mannering's state of mind after being consumed is far from vexation; rather, it is one of perplexed acceptance of the entire situation.

"Mr. Sycamore" takes the state of vegetable torpor one step further and offers readers an existential twist to combat the problems of life as the main character, John Gwilt, resolves to become a tree. Although most of the story is presented comically – all John seemingly manages to do for a while is catch a cold – there are serious moments of contemplation concerning human existence. When John is questioned by the local newspaper, he responds:

> I told you I loved life. I want to be alive, really alive; I want to live in the earth's

> way. We are all estranged. Do you understand what I mean? We are broken off from the earth. I want to be part of the earth's life. I am tired of this helpless fluttering, all this hurrying and scurrying: it isn't dignified; it isn't sane; I want to stand still … I'll go on living, for a hundred years – a thousand! […] Never suffering the pains and miseries of growing old; enjoying the sun and the wind and the rain, too, for a change; the blue sky in the morning and the twinkling stars at night; watching the world go by. (107-8)

John's fascinating understanding of nature's existence as being wholly better than human existence makes plant life seem much more favorable and desirable than much of today's killer plant narratives do. Still, none of these pieces of literature has complicated the relationship between man and plant more dangerously or problematically than *The Little Shop of Horrors*.

Scholars in various fields of study have explored *The Little Shop of Horrors*' deeply-rooted ideological messages. Mikita Brottman, for instance, references the movie, arguing that, "unlike animals, robots, and zombies, plants don't turn on their owners, they don't hold grudges, and they can't gang up and plot revenge. In fact, there's something ridiculous in the idea of being terrorized by a plant, which is why the subject makes such fertile ground for spoofs like *The Little Shop of Horrors* (1960) and *Attack of the Killer Tomatoes*! (1978)." Charlotte Greenspan offers a more specific reading of the story in terms of the function of death, arguing in one example that *Little Shop of Horrors* is "drenched in death" (158), since Audrey II has killed almost all of the principal cast members by the end of the stage adaptation. Alternatively, Jensen focuses his analysis of *The Little Shop of Horrors* on the musical versions of the story and makes several key arguments centering primarily on social issues of race and class. Jensen asserts that Audrey II's victims are characters who would otherwise be obstructing Seymour's upward social mobility and that there is strong association of Audrey II to African American influences, including speech, song, and form (54). Although several scholars thus far have offered important arguments concerning the functions of race, class, death, music, and even environmentalism in *The Little Shop of Horrors*, the symbiotic relationship existing between the film's human sphere, mainly represented by protagonist Seymour Krelboyne, and the carnivorous plant Audrey Junior has yet to be fully explored.

Killer Plant Turned Cult Classic. The Little Shop of Horrors is the tale of Seymour Krelboyne, an aspiring botanist who crossbreeds Audrey Junior, a plant which turns out to be carnivorous, with a specific appetite for human blood and body parts. The original film takes place in Los Angeles' Skid Row neighborhood and Mushnick's Florist, the local flower shop where Seymour works. Although the film accentuates a romantic storyline between Seymour and his female love interest, Audrey, after whom the plant is named, much of the film revolves around Seymour's attempts to satisfy Audrey Junior's hunger without anyone else finding out what the plant's food of choice really is. By problematizing the relationship between man and nature, the uniqueness of *The Little Shop of Horrors*, compared to other killer plant narratives, lies precisely in its nuanced treatment of the so-called killer plant.

Since its original black-and-white release on the big screen, Roger Corman's 1960 film, written by Charles B. Griffith, has gone through two major re-vamps. In 1982, composer Alan Menken and writer Howard Ashman teamed up to create a musical based on the original film. Then, in 1986, director Frank Oz led a film adaptation of the musical, and this last version, starring Rick Moranis, has, since its release, become the better known version of the famous tale. There are, of course, several nuanced differences between the 1960 and 1986 versions: the spelling of Seymour and Mushnick's last names change (Krelboyne to Krelborn, Mushnick to Musnik), the name of the killer plant changes (Audrey Junior to Audrey II), and the sadistic Dr. Farb becomes Audrey's abusive – but still sadistic – boyfriend Orin. The most noteworthy change which affects the story's ideological messages, however, is that the endings of all three versions differ dramatically, a point that will be explored in detail. Additionally, although the following section focuses on the original 1960 film, I will also address the significant changes between the movie and the Broadway versions of the story, modifications which have a substantial effect on the socio-political tone and transcendental (or anti-transcendental) implications of the original text itself. Despite all of these discrepancies, Audrey Junior and Seymour are, in all versions, shown to share in an interdependent relationship, where one cannot thrive without the other. *The Little Shop of Horrors* thus stands out from other killer plant movies in its return to a symbiotic relationship between man and plant.

Seymour Krelboyne and Audrey Junior: Friends or Foes?

The 1960 version of the relationship between Seymour and Audrey Junior is much more amiable than the later relationships. First, Audrey Junior, unlike its musical counterparts, is not an alien plant from outer space but simply a consequence of Seymour's crossbreeding[3]. Necessarily, then, Seymour's relationship to the plant is that of primary caretaker, responsible for nursing it[4] back to health. Seymour is accordingly positioned as a friend to Audrey Junior, and vice versa, rather than as a foe. Audrey Junior, in return, provides Seymour psychological rewards. Seymour's emotional transition from beginning to end is the most drastic of all the human characters. From the onset, Seymour is portrayed as nerdy, weak, unintelligent, and unsure of himself. He still lives at home with his mom and is exceptionally clumsy. However, as soon as Audrey Junior begins to grow, Seymour immediately becomes the talk of the town and gains a remarkably high social status within the community; even the "Society of Silent Flower Observers of Southern California" visits the store to reward him with their annual trophy for creating such a "magnificent bloom." As the story moves forward, Seymour becomes more confident in himself; he gets the girl of his dreams and believes that his recognition with the prestigious botany club could bring him fame and fortune.

Mr. Mushnick, too, begins as a grumpy, unhappy, and low-income character but eventually undergoes a positive transition. Because of Audrey Junior's fame, Mushnick eventually becomes giving, happy, and rich, and as Audrey Junior thrives, so does Mushnick's business. He gleefully gives Mrs. Shiva, a regular shop customer, a free bouquet of carnations and also offers Seymour a pay raise. More importantly, Mushnick begins feeling warmly towards Seymour, putting an arm

---

3 Audrey II's "alienness" is, however, heavily emphasized in the musical adaptation, specifically in Seymour's retelling of how he finds the plant in the song "Da-Doo" and Audrey II's explanation of its own origins in the song "Mean Green Mother from Outerspace."

4 Although the plant is named after Seymour's romantic love-interest, a female character, the plant's rumbling voice would suggest a male identity, as well as the "Junior" or "II" generational suffixes normally apply to male lineage as opposed to female lineage. In addition, men were always cast as the voice of Audrey Junior/II – writer Charles B. Griffith in the 1960 film, Ron Taylor in the original 1982 Off-Broadway cast, and Motown singer Levi Stubbs in the 1986 film adaptation. In the case of pronoun use, however, characters in the films generally refer to Audrey Junior/II as "it," instead of he or she.

around him and proclaiming, "Look Audrey, I got a son […] I want you should call me dad." Mushnick, moreover, reaps all of the economic blessings that come with such a magnificent, unusual plant. At the beginning of the film, Mushnick proclaims, "I can't even sell the plants I got in my shop!", implying that his business is slow and struggling to stay afloat financially. However, his flower shop immediately begins to thrive because of Audrey Junior and also gains prominence in the otherwise demoralized location of Skid Row: a multitude of customers begin visiting the shop to see the plant, which brings about an increase in sales. Specifically, two girls from the local high school arrive and pay Mushnick $2000 to supply flowers for their Rose Parade float.

Essentially, Audrey Junior's overall presence enhances the lives of the main characters as well as the economic situation of Mushnick's Florist and of the Skid Row community. Its presence revitalizes the economic and emotional conditions of both the store and the characters involved, as well as transforms, however briefly, the relationship between Mushnick and Seymour from employer and employee to father and son. Additionally, Audrey Junior attracts all types of people from various cultural institutions: the local high school, the nationally-recognized Rose Parade, and the Society of Silent Flower Observers of Southern California, customers that Mushnick's store and the community would not ordinarily draw. The social conditions pre- and post-contact with Audrey Junior displays, ultimately, the subtle, sympathetic attitude towards plants being shaped by the film.

Death in *The Little Shop of Horrors*: Symbolic and Symbiotic

Audrey Junior consumes a total of five human bodies: the man at the railroad tracks, Dr. Farb the dentist, the armed robber at Mushnick's store, Leonora Clyde, and Seymour. By looking at exactly who dies, how they die, and what each death symbolizes, as well as investigating what the scenarios all mean in light of the eventual death of Seymour, the notion of death in *The Little Shop of Horrors* becomes less about a killer plant's "evil" motives and more about Seymour and Audrey Junior's mutually-beneficial relationship.

While some may view the characters who die at the hands – or pitchers, rather – of Audrey Junior as innocent bystanders, a look at the societal role that each character embodies suggests a more complex understanding of the death-by-plant scenario. The man at the railroad tracks, before getting hit by a large rock Seymour tosses in his direction, seemingly wakes up from behind a leftover wagon and reaches for a bottle of bourbon. This man, while probably not part of the homeless Skid Row demographic based on his seemingly neat clothing, may have been an alcoholic and would, then, not be considered an asset to the community. Similarly, the armed robber who attempts to steal money from Mushnick's flower shop and Leonora Clyde, the aggressive female hooker who continuously attempts to gain the attention of a hypnotized Seymour, would also not be considered beneficial to the community; in fact, in their most basic forms, a criminal and a prostitute would be considered a corrupting component of the community. Such careers are looked down upon in most societies and such individuals are regarded as selfish, looking out only for themselves and performing actions for their own financial gains. Dr. Farb, on the other hand, seemingly presents a different aspect of the community, since he is the local dentist and would, ostensibly, be a respected figure in the neighborhood. However, his overtly sadistic nature denotes otherwise. He enjoys inflicting pain on his patients, unnecessarily pulling out teeth, and he arguably has bad medical practices (except, of course, you ask his masochistic patient, Wilbur Force). His death, then, like that of the others, is not necessarily considered detrimental for the community; on the contrary, Audrey Junior could be understood as helping to eliminate the "bad eggs" of society.

Moreover, many of the deaths demonstrate more about Seymour's character, specifically his perpetual clumsiness and bumbling ways, than they do about any hint of aggression Audrey Junior might have towards human life. The way in which the deaths occur reveals, as well, the larger function of Audrey Junior. The rock Seymour throws at the tracks was not meant for the man, who, after being hit, drunkenly stumbles over the tracks while Seymour merely panics, running back and forth; Seymour's only effort to help is to bring a bucket of water, which, of course, he spills before it can be of any use. In addition, even though Seymour is hypnotized and on the hunt "to find food for master," Leonora Clyde's death is an unintentional murder, actually caused by Seymour's "flipping" of a rock. Farb's death, while not out of clumsiness, occurs in blundering self-defense because Seymour believes that Farb is attempting to kill him during his dental appointment. Likewise, Mushnick, attempting to avert the attention of the armed robber, tricks the criminal into climbing into the pitchers of

Audrey Junior, and this can also be considered an action of self-defense.

It is, perhaps, even more significant to note that only two of these deaths actually happen at the pitchers of Audrey Junior. Both the armed robber, however deceived by Mushnick, and Seymour voluntarily climb into the plant and are then eaten alive. However, the man at the tracks, Dr. Farb, and Leonora Clyde all die outside of the flower shop and by the hands of Seymour. The man on the tracks and Leonora Clyde are both unintentionally killed by Seymour's bad rock tosses and Farb is killed in a moment of self-defense. Audrey Junior, then, is actually "saving" Seymour by covering up the deaths that Seymour would otherwise be arrested for.

The musical versions, on the other hand, offer much more plant-antagonistic endings than the 1960 film. Remember that the five "deaths" in the original film are all minor characters except in the case of Seymour himself: the man at tracks, Dr Farb, the armed robber, and Leonora Clyde. In addition, the final scene shows Audrey Junior as dying along with the humans; the final bud revealing Seymour's face droops over, and, because there was no offspring of Audrey Junior's, the audience can safely assume that this is the end of the plant's saga. However, in the stage musical, Audrey II claims the lives of all four main characters: Orin, Mushnik, Audrey, and Seymour; furthermore, audiences are told at the end that Audrey II has continued tricking people for their blood in exchange for fame and fortune. Quite similarly, in the musical film, although Audrey II only manages to consume Orin and Mushnik, a mini-Audrey II is seen smiling, growing in the garden of Seymour and Audrey's new house.[5]

Not only, then, do these musical adaptations make the plant the sole murderer of humans, but they also, perhaps wrongly so, heighten the tension between man and plant by presenting the aggressive, devilish desires of Audrey II, who attempts to taunt and tempt Seymour with success, money, and popularity. The deaths that occur in the musical adaptation happen out of spite: Audrey II presents Orin to Sey-

---

[5] Making things even more problematic is what Frank Oz, who directed the musical film adaptation, and screenwriter Howard Ashman claim to be their original intentions for their 1986 adaptation. Even though the 1986 ending is not as devastating as the stage musical, Oz explains in a 2012 interview that they had always wanted the two main characters, Seymour and Audrey, to be eaten by Audrey II and for Audrey II to live; however, the original twenty-three minute ending had to be scrapped, due to the very negative reception of two different test audiences.

mour as an individual who should be killed because of his abusive actions towards Audrey, and Mushnick holds Seymour at gunpoint and threatens him. Most significantly, these musical adaptations allow Audrey II's reign to continue, making the tone of the story much more apocalyptic. Instead of the portrayal of a symbiotic relationship between man and plant, there is a separation and distancing in the relationship, and the adaptations actually revert back to the more typical killer plant narrative ideology.

Seymour Krelboyne: A Model for Plant-Sympathy

While the original film does show a sense of anxiety towards plants – as evidenced by Audrey Junior's ability to put Seymour into a hypnotic state, for instance – and, consequently, a possible fear that plants may be able to manipulate the human mind in some way, the original movie's overall attitude towards plants is much more favorable and accepting of Audrey Junior, as opposed to the clear pitting of man against plant in other killer plant narratives. In fact, the 1960 film depicts a distinct, symbiotic relationship between Seymour and Audrey Junior, where both benefit from the other. Seymour and Mushnick, additionally, do not fear being eaten by Audrey Junior themselves; rather, both are eventually seen as standing up to it, yelling back at it when it demands food. Significantly, Audrey Junior is seen as a way out, or form of escape, for Seymour, rather than a taunting, scary monster.

Thus, Seymour becomes a prototype for human plant-sympathy. In his memorable speech to Audrey Junior, Seymour frustratingly asks, "Who raised you from a bunch of little seeds? Who fed you all them high-class fertilizers and sat up all night with you when you were sick? Nobody else woulda done that for you. You think anyone else would've brought you human beings to eat? You're darn right they wouldn't. Well, I've helped you, and you've helped me." Audrey Junior and Seymour, as symbiotic relationships normally function, ostensibly need each other to survive. They seem, in fact, to "save" each other and sacrifice for one another throughout the film. Seymour pricks himself and sacrifices the blood in all ten fingers to help sustain Audrey Junior, and he later brings corpses for the plant to eat. Audrey Junior, in return, covers up the deaths of those that Seymour accidentally kills. And, in the end, when the cops are finally on to Seymour's murders and are after him, Audrey Junior can be thought of as covering up Seymour, quite literally, in the final scene. Finally, the death of Seymour brings about the death of Audrey Junior as well, exemplifying the exceedingly entangled relationship between man and plant.

Growing Green in the West: Signaling a Movement Back to Transcendentalism

> I believe that there is a subtle magnetism in Nature, which, if we unconsciously yield to it, will direct us aright. (Thoreau)

One final point to address is the coinciding values and chronology of the Green Movement with the ideology and release of *The Little Shop of Horrors*. While the roots of the Green Movement can be found in the Transcendentalist Movement and the Progressive Era, which thrived in the 1800s and 1900s respectively, the modern Green Movement did not garner much attention or support until the 1960s. Additionally, Transcendentalist leaders like Thoreau and Emerson argued that society corrupted the individual, asking, instead, that individuals seek an understanding of themselves in relation to nature and the universe; World War II sparked concern regarding the physical effects the war had on the Earth's natural resources, and, today, the concerns of environmental campaigns in America focus on human efforts to combat issues like the overflowing of landfills, anti-nuclear technologies, climate change, and global warming. In view of this, the latter part of the 1950s brought about a multitude of killer plant narratives, as I discussed earlier, and this outbreak of man versus plant scenarios in mass media may be society's attempt to negotiate an understanding of human-plant relationships. The production period of *The Little Shop of Horrors* around 1960 certainly overlaps with the time period of the society's growing attention for the modern Green Movement. And, if we consider *The Little Shop of Horrors* in light of Seymour and Audrey Junior's symbiotic relationship, the film would certainly fit in with Transcendental values and the Green Movement's call for action to understand human-plant contact.

In more recent years, the representation of apocalyptic, environmental catastrophes in mass media has been criticized for its man-on-top ideology. Kirkpatrick Sale, an environmentalism scholar, contends that, "[e]ver since Hiroshima began our knowledge of environmental disaster we have produced movies, novels, TV shows and suchlike that show what environmental crises, even a worldwide catastrophe, would look like. Note, of course, that humans always survive these crises" (26). Killer plant narratives, however, present an even darker catastrophe: the plant has begun retaliating towards human life, and humans may not always be the last to survive. Brottman also considers how a closer examination of ideologies revealed in killer plant narratives renders the films more serious than spoof and may also be telling of America's ecological situation:

> [P]erhaps these nightmares of nature are a projection of our own green guilt. We devastate foliage with pesticides and bring forests to the brink of destruction; we genetically deform vegetables into monstrous hybrids and cause irreparable damage with weed killers and growth hormones. What's more, we cultivate plants solely in order to eat them, or to cut them off in their prime and enjoy the way they smell as they slowly die. With no Parliament of Trees to defend their rights, perhaps we can't help but unconsciously imagine plants to be, at some level, voiceless, resentful, and oppressed. That may be the anxiety at the root of films with smart, vengeful plants cutting humans down in our prime. We reap what we sow. (B12)

Looking at killer plant narratives in this way, the apocalyptic, terror-ridden stories can be understood functioning as Brottman suggests: to show how badly we have been treating the world in terms of ignorantly and destructively using up natural resources. Killer plant narratives are a visual illustration and understanding of what nature would do and say to humans if they could react to our adverse actions.

The Little Shop of Horror stands out from the rest of the killer plant crowd, however, in its understanding of how man and plant can work together in a harmony, albeit a delicate one, instead of acting out the traditional doomsday scenario. Seymour's symbiotic relationship with his plant, Audrey Junior, depicts the relationship man should have had with nature all along; man as a responsible caretaker for the earth and plant life. Whether Seymour wanted it or not, he has become entangled with Audrey Junior; each entity necessarily needs the other in order to survive, and this relationship is the 1960s' rough sketch of how humankind should interact with Mother Earth.

Archetypal killer plant narratives can be understood as revealing the effects that our hunger and greediness has on Earth's natural resources, especially during a time in which the Western world is consumed by all things fast and cheap. Killer plant narratives, as with many horror and apocalyptic storylines, function as warnings for what may be lurking ahead and also as attempts to negotiate and understand what is going on more globally. Thus, as we move further along into the twenty-first century, we must consider the specific ecological conditions that each nature narrative presents to audiences.

Certainly, plant-sympathetic storylines like the original *Little Shop of Horrors* stand out from the rest of the crowd and teach us that humans can and should consider plants as our equals. *The Little Shop of Horrors*, and plant-sympathetic tales like it, revolutionizes the typical killer plant ideology of the man-on-top, socio-ecological hierarchy, heartily problematizing it in every way. Instead of representing unconscious American fears and the anticipation of ecologically- and environmentally-detrimental situations to come like many of its killer plant story counterparts, the symbiotic relationship depicted in *The Little Shop of Horrors* represents the mutually-beneficial relationship that man and plant can have in the real world, an optimistic invitation for society to function with nature instead of against it.

Works Cited

Attack of the Killer Tomatoes. Dir. John De Bello. Perf. Jerry Anderson, Don Birch, Tom Coleman, Art K. Koustik, Ernie Meyers. Rhino Home Video, 2003. DVD.

Ayre, Robert. "Mr. Sycamore." *Man into Beast: Strange Tales of Transformation*. Ed. A.C. Spectorsky. Garden City: Doubleday, 1947. 89-149. Print.

Brottman, Mikita. "When Good Plants Go Bad." *Chronicle Of Higher Education* 55.16 (2008): B12-B13. *Academic Search Elite*. Web. 15 Jan. 2013.

Collier, John. "Green Thoughts." *Man into Beast: Strange Tales of Transformation*. Ed. A.C. Spectorsky. Garden City: Doubleday, 1947. 71-88. Print.

Creepshow. Dir. George A. Romero. Perf. Stephen King. Warner Home Video, 2009. DVD.

The Day of the Triffids. Dir. Steve Sekely. Perf. Howard Keel. Allied Artists, 1998. VHS.

Dr. Terror's House of Horrors. Dir. Freddie Francis. Perf. Christopher

Lee, Ann Bell. Paramount Pictures, 1999. DVD.

Emerson, Ralph Waldo. "Nature." *The Norton Anthology of American Literature*. 7th ed. Ed. Nina Baym. New York: W.W. Norton & Company, 2008. 492-519. Print.

Gould, Stephen J. *Leonardo's Mountain of Clams and the Diet of Worms: Essays on Natural History*. New York: Harmony Books, 1998. Print.

Greenspan, Charlotte. "Death Comes to the Broadway Musical." *Daedalus* 141.1 (2012): 154-9. *EBSCO*. Web. 15 Jan. 2013.

The Guardian. Dir. William Friedkin. Perf. Jenny Seagrove, Dwier Brown. Anchor Bay Entertainment, 1999. DVD.

The Happening. Dir. M. Night Shyamalan. Perf. Mark Wahlberg, Zooey Deshchanel, John Leguizamo, and Betty Buckley. 20th Century Fox, 2008. DVD.

Invasion of the Body Snatchers. Dir. Don Siegel. Perf. Kevin McCarthy, Dana Wynter. Olive Films, 2012. DVD.

Jensen, Marc. "'Feed Me!'": Power Struggles And The Portrayal Of Race In Little Shop Of Horrors." *Cinema Journal* 48.1 (2008): 51-67. *Academic Search Elite*. Web. 20 Dec. 2012.

Ley, Willy. *Salamanders and Other Wonders: Still More Adventures of a Romantic Naturalist*. New York: Viking Press, 1955. Print.

Lipske, Michael, and Robert Noona. "Forget Hollywood: These Bloodthirsty Beauties Are For Real." *Smithsonian* 23.9 (1992): 48. *Academic Search Elite*. Web. 15 Jan. 2013.

Little Shop of Horrors. Dir. Frank Oz. Perf. Rick Moranis, Ellen Greene, Vincent Gardenia, Levi Stubbs, Steve Martin. Warner Home Video, 2009. DVD.

—. Dir. Roger Corman. Perf. Jonathan Haze, Jackie Joseph, Mel Welles, Jack Nicholson, Dick Miller, Charles B. Griffith. Legend Films, 2008. DVD.

Minority Report. Dir. Steven Spielberg. Perf. Tom Cruise. 20th Century Fox, 2002. DVD.

Osborn, Chase S. *Madagascar, Land of the Man-Eating Tree*. New York: Republic Publishing Company, 1924. Print.

Oz, Frank. Interview by James Gartler. "Frank Oz: Muppets maestro discusses 'Little Shop of Horrors' and the remaking of his classics." *Entertainment Weekly*. Entertainment Weekly Inc., 2012. Web. 15 Aug. 2012.

The Ruins. Dir. Carter Smith. Perf. Jonathan Tucker, Jena Malone.

Paramount Pictures, 2008. DVD.

"The Seeds of Doom." *Doctor Who (Season 13)*. Dir. Douglas Camfield, David Maloney, Paddy Russell, Barry Letts, Christopher Barry. Perf. Tom Baker, Elisabeth Sladen, John Gleeson, Michael McStay. BBC, 2011. DVD.

The Thing from Another World. Dir. Christian Nyby. Perf. Margaret Sheridan, Kenneth Tobey, Robert O. Cornthwaite, Douglas Spencer. Turner Home Entertainment, 2003. DVD.

Thompson, D'Arcy Wentworth. *On Growth and Form*. New York: Dover Publications, 1992. Print.

Thoreau, Henry David. "Walking." *The Atlantic*. The Atlantic Monthly Group, 1 Jun. 1862. Web. 12 Jan. 2012.

"Top 10 Killer Plants." *Den of Geek US*. Dennis Publishing, 6 Aug. 2008. Web. 2 Aug. 2012.

Mean Green Machine: How the Ecological Politics of Alan Moore's Reimagination of Swamp Thing Brought Eco-consciousness to Comics

Hindi Krinsky

Abstract: This essay examines how Alan Moore's Swamp Thing series introduced ecological awareness to the world of comics. Prior to Moore, nature was typically depicted as a one-dimensional space mainly to provide narrative texture. Under Moore's tutelage, the environment became a lead character. By reconfiguring Swamp Thing as an all-vegetative entity, Moore forced the reader to commune directly with nature itself. This essay explores Moore's ecological vision along with the environmental movements of the time. Looking beyond this body of work, it concludes with a brief discussion on Moore's eco-legacy and impact on contemporary comics.

Though comic books began appearing in 1933, a seismic shift occurred in 1938 with the debut of Superman in *Action Comics #1*. This iconic character not only heralded a new genre of heroic storytelling, but set the standard, tone, and aesthetic for decades of comics. In the 1930s, Superman was presented as a kind of everyman, a powerful vector of wish fulfillment for an economically depressed nation. With his relatable alter-ego Clark Kent and immigrant backstory (as a refugee from the planet Krypton), Superman also challenged the boundaries of social acceptance and spoke to the many newly-minted Americans of the 1930s and 40s (Pevey 16). Published throughout World War II, Superman soon became symbolic of the war efforts and helped popularize the selling of war bonds (Daniels 64). In short, Superman became the driving force behind the popularization of comics books and by the end of the 1940s, he was joined by a pantheon of superheroes, including Batman, Wonder Woman, and Captain America, characters that collectively defined the ethos of American comics.

During the golden age of the industry,[1] new missions, powers, and identities (Coogan 82) were constantly developed. Some characters like Superman had otherworldly abilities which derived from their extraterrestrial origins while others like Aquaman were human-hybrids and possessed more earthly talents. With this onslaught of superheroes and their rich fantasy worlds, adults grew increasingly concerned that comics promoted corruptive morals and provided a warped view of reality, causing rebelliousness, delinquency, illiteracy, and even suicide among young readers.[2] Fearful of being shut down by Congress or worse, publishers of comics committed to self-regulate their material and formed the Comics Code of Authority (CCA). The code, enacted by the Comics Magazine Association of America, censored texts for violent language, nudity, and most importantly, ensured that government officials always "triumphed over evil [...] [and] were not depicted in a way to create disrespect for established authority" (Comics Magazine Association of America par. 2). Though this move allowed most publishers to retain some control over their material, comics, with the exception of MAD (which converted to a new format to avoid the CCA[3]) "lost their social relevance" (Weiner 8).

As the industry evolved into its silver and bronze ages, most publishers began openly rejecting the censorship of the CCA, becoming decisively more relevant and issue-oriented. This content drew adult fans – many of whom had been raised on a steady diet of comic books. Alan Moore's *Swamp Thing* was one of the first to disregard the CCA and feature more mature themes, conveying a sophisticated portrait of the environment within a classic horror story. In presenting Moore's ecological vision, this essay will discuss how the environment was historically employed in early comics, specifically as narrative backdrop or as sacred space. It will then discuss how these conceptualizations were confronted head-on in Moore's work which echoed the environmentalism movements of the 1970s. Looking beyond this body of work, this essay will conclude with a brief discussion on Moore's eco-legacy within the world of comics.

---

[1] According to most scholars, the golden age of comics began with the publication of Superman in 1938 and lasted to the early 1950s. See John Carlin, *Masters of American Comics*. New Haven: Yale University Press, 2005.

[2] These fears were promulgated by psychiatrist Fredric Wertham whose 1954 book, *Seduction of the Innocent,* addressed issues of violence, sex, and drugs in comics.

[3] See Maria Reidelbach, *Completely MAD: A History of the Comic Book and Magazine.* Boston: Little, Brown, 1991.

Conceptualizations of Nature in Early Superhero Comics

Aquaman first appeared in *More Fun Comics #73* in 1941 as a backup feature in DC's anthology title *Dr. Fate*.[4] In this issue, Aquaman is introduced as a sea-legend, the "most amazing of all the secrets of the sea […] who, like a water-god of old, dwells in the dim unknown depths of the ocean, leaping forth to do battle with the evils and injustices of the world we know" (1). On the splash page, Aquaman is costumed in a golden scaly chain-mail-like tunic with green gloves and leggings, protecting a mother and child from two assault-weapon-wielding Nazis. In this 8-page feature, the reader is introduced to Aquaman as he emerges from his watery kingdom to save a ship of refugees as they are pursued by a Nazi U-boat. In a dramatic sequence, "at the very instant the gun belches forth its shell- a hand from out of the sea!" (2). Aquaman chases away the submarine, calling himself "[a] metal fish that strikes like a killer shark" (3). In the rescue, Aquaman enlists the help of his below-the-sea friends, addressing a pod of porpoises "in their own language" (3) and urging them to move the small lifeboat of refugees close to land. When the captain expresses his gratitude and asks him his name and nationality, Aquaman replies "[I] am from <u>no</u> land- my name is Aquaman" (3). When pressed further, Aquaman explains,

> The story must start with my father, a famous undersea explorer—if I spoke his name you would recognize it. My mother died when I was a baby, and he turned to his work of solving the ocean's secrets. His greatest discovery was an ancient city, in the depths of where no other diver had ever penetrated. My father believed it was the lost kingdom of Atlantis. He made himself a water-tight home in one of the palaces and lives there, studying the records and devices of the race's marvelous wisdom. From the books and records, he learned ways of teaching me to live under the ocean, drawing oxygen from the water and using all the power of the sea to make me wonderful strong and swift. By training and a hundred scientific secrets, I became what you see – a human who lives and thrives under the water. (3-4)

[4] In comics, popular characters often go through multiple evolutions. This essay only addresses the very first Aquaman feature which may be inconsistent with later versions of the character. In essence, the later versions of Aquaman – as a member of the Justice League, Arthur Curry, or the Atlantean Orin – are different characters.

Following this exchange, Aquaman – with the help of his aquatic friends – gives chase to the U-boat, commenting "there is much evil in this upper world. I will do my share in punishing it. Now- to settle accounts!" (4). Though the Nazis attempt to drown Aquaman in a high-pressured volcano, the loyal sea creatures quickly come to his aid. Emerging from the water, Aquaman returns to fight the Nazis. When one comments "I must escape the sea-monster," Aquaman retorts "You land-monsters are far worse" (8). The Nazis' island base is subsequently destroyed and, though Aquaman returns to his underwater kingdom, the reader is assured that "from his lair in the deep he rises to face new dangers and win new triumphs" (8).

In this premier feature, Aquaman, though born of human parents, is reborn by the sea. The oceans where Aquaman is raised are portrayed as being removed from "the upper world" (4), away from terrestrial humanity, a primordial womb that rebirths him as a superior being. In this place, where his human father figuratively 'drowns' himself in grief/scientific research following the early death of his mother, the sea assumes the role of motherhood,[5] echoing an ancient Homeric literary tradition where the "ocean [is] fountainhead of the gods, and Mother Tethys who nourished me in their halls and reared me well" (14.201). His identity is so thoroughly molded by the sea that his father calls him "a true dweller of the deep" (4) and renames him Aquaman. In this way, the sea-world is recast as a sacred space, a font of baptismal waters in which a new god is raised in isolation to receive mysterious training and develop herculean abilities. Yet, though he is called a god, Aquaman does not seem to have full dominion over the sea – he cannot escape the underwater volcano alone or mobilize the plankton to help the refugees. He is skilled in the ways of the sea (communicating with visible creatures) but is not of the sea itself. Instead Aquaman appears to be a member of the ocean community – not its ruler – and is therein subject to its natural laws and ecology.[6]

Though water is often symbolic of purification and rebirth, it is also conceptualized as a source of destruction and death (Ferber 180). Similarly, Aquaman is portrayed as an enforcer, emerging from him womb-like sanctuary to mete out justice and restore balance to the world. Like Noah's biblical flood, Aquaman seems to arise from a

[5] Aquaman is often caricaturized as an effeminate superhero. This feminization could be contributed to his mother-childlike relationship with the sea.

[6] In more modern version of Aquaman, he is the king of Atlantis.

disaffected space to assist the good and punish the wicked. The text offers no explanation as to why he suddenly emerges and intercedes on behalf of these particular refugees. It is also unclear why he retreats undersea after destroying a single Nazi U-boat and island outpost. Presumably, the lower-world exists at a distance from terrestrial life and is undisturbed by the genocide above-ground. This suggests that Aquaman, though reared by the sea and brotherly to many sea-creatures, is simply a mediator between the aquatic and terrestrial worlds, acting as an occasional prophet or messenger who utilizes natural resources in fulfilling his personal missions.

With his liminal status, Aquaman is positioned between two parallel earths, between distinct upper and lower worlds. The upper-world – terrestrial life – is the sphere of humans, as "people are home on the land" (Ferber 179). For this reason it is a place that seems to inhabit and possess evil. In contrast, the lower world – life beneath the sea – appears paradoxically simplistic and mysterious. On one hand, Aquaman has no difficulty convincing sea-creatures to assist him in escaping the underwater volcano. On the other hand, nature as a force neither impedes nor assists in his rescue efforts. The lower-world is an enigma, unknown, so far removed from humanity that even the lost city of Atlantis, Aquaman's home, is deserted and uninhabited by people save for him. This suggests that the sea, in this case representing an unmolested state of nature, is mostly undisturbed and unmoved by the actions of humans, existing in a kind of "pristine" or "wild" state. As it is cast in this sacred light, it also appears somewhat one-dimensional.

Though this essay only addresses the very first Aquaman feature, it suggests that this early work largely ignored the interdependence of humans and nature, presenting instead a simplified though imaginative view of life beneath the sea. With its color-saturated coral reefs and dramatic aquatic creatures, this comic introduced a new exotic locale, a rich landscape in which to set exciting adventures stories, but not a forum to address environmental issues. In conveying nature as a distant and unmoving entity, this early comic suggests a static and disaffected environment that seems uninhabited and even disinterested in humankind. Aquaman, a human reared underwater, coexists with sea-life while remaining somewhat involved in the terrestrial world as well – an interest spurred by his humanness, not by the sea itself.

In some ways, the Aquaman narrative echoes the environmental attitudes and natural philosophy of American conservationist John Muir. At the age of thirty-eight, Muir retreated into nature to make his home in the primordial wilderness, "to follow where his communion with nature led him" (Austin 12). In the solitude of the forests, Muir committed himself to communicating with his natural companions, to "interpret the rocks, learn the language of flood, storm, and the avalanche [to] get as near the heart of the world as I can" (Austin 13). Like Aquaman, Muir's identity is not only crystallized in the wilderness, but flowers into the recognition that he is a subject of nature itself. In a letter to his friend Jeanine Carr, Muir describes this commitment:

> Do behold the King in his glory, King Sequoia! Behold! Behold! Seems all I can say. Some time ago I left all for Sequoia and have been and am at his feet, fasting and praying for light, for is he not the greatest light in the woods, the world?...But I'm in the woods, woods, woods, and they are in *me-ee-ee*. The King tree and I have sworn eternal love – sworn it without swearing, and I've taken the sacrament with Douglas squirrel, drunk Sequoia wine, Sequoia blood, and with its rosy purple drops I am writing this woody gospel letter. (Oates 122)

This Christ-like imagery speaks to Muir's philosophy that nature is an enigmatic and sacred space, a place that is "pure and good…that was separate from God, but [where] each organism existed to glorify God" (Worster 8). In this way, nature is framed as something far removed from people, a heavenly sphere untouched by the artifice of human life. This strongly resembles Aquaman's own vision of dichotomous upper and lower worlds, where evil monsters exist on land destroying one another purposelessly. Like Aquaman's unaffected lower-world, Muir conceptualized nature as "the name given to the part of the world that we humans did not create, that we do not manage, and that can survive our extinction" (Williams 8). In other words, nature in its elemental wildness is not truly affected by the trials and tribulations of humankind, but exists both beyond and outside of it.

Like Aquaman, Muir existed in a liminal space, occasionally emerging from nature to participate in human life. Oates argues that Muir "struggled all his life to balance his need for ordinary human sustenance and connectedness with his desire for what he habitually labeled 'pure' wilderness […] [considering] them as irreconcilable opposites" (122). Like the Aquaman enforcer, Muir came to be synonymous with the natural world, becoming a kind of messenger of the wilderness, a natural prophet that preached that "there is more to the world than humankind and its artifacts" (Worster 9). Similarly, Aquaman exposed comics readers to a new natural landscape, a frontier beyond human experience that was both unspoiled by and inaccessible to its readers.

Swamp Thing & Moore's Ecological Vision

The Swamp Thing character was created by writer/artist duo Len Wein and Bernie Wrightson. It originally appeared in 1971 as a non-serial turn-of-the-century horror story in DC Comics' anthology series, *House of Secrets.* In this short feature, Swamp Thing is Alex Olsen, a brilliant Louisiana-based scientist who is nearly killed by his jealous lab partner, Damien Ridge. Though the bomb explosion destroys his lab and critically wounds him, Olsen manages to escape to the nearby bayou where his consciousness seeps into the boggy soil and emerges as a human-swamp-hybrid, a dark mud creature with human awareness, but no discernible voice. Later, when Damien attempts to kill Olsen's wife, Linda, Alex/Swamp Thing kills his traitorous colleague and saves her. When he tries to communicate with Linda, she is frightened by her mud-dripping rescuer and does not recognize him as her loving husband:

> The tortured, shattered look in her once-sparkling eyes is more than I can endure – I turn my face away from her – and I start to go home. Only the swamp is kind to me – It is only the swamp that cares [...] if tears could come – they would! (Wein 1: 8)

Broken-hearted, Swamp Thing returns to his new home, a dark lonely habitat where he is overwhelmed by a sense of otherness, typecast as a displaced monster – alienated and powerful, but belonging nowhere, not in nature or civilization.

Recognizing the potential for a new long-term DC character, Swamp Thing was soon afforded his own title series. This time, the character was configured as Alec Holland, a brilliant biologist based in the contemporary (1970s) Louisiana. In his top-secret lab, Alec, employed by the US government, works on the development of a bio-restorative formula that can "make forests out of deserts" (Wein 1:

16). When the villainous Mr. E. is unable to convince Alec and Linda to forfeit their research, he plants a bomb in the lab which critically injures Alec. Engulfed in flames and splashed by his formula, Alec escapes to the swamps where he fuses with the local plant life and emerges as the Swamp Thing. The original ambling mud-mute character is replaced by a muscular, verdant, and uncommunicative superfigure. Presumed dead by his friends, Mr. E.'s henchmen later return and murder Linda, sending Alec/Swamp Thing into a rage. In the subsequent twenty-three issues, Swamp Thing avenges Linda's death, wages war against his arch-nemesis, the insane Dr. Anton Arcane, and develops a close relationship with Arcane's niece, Abby. In this evolution of the character, Swamp Thing, much like Aquaman, identifies himself as a human-hybrid: a man in essence, but with the power to harness nature in unnatural ways. Unlike Aquaman, who seems at home in his new habitat, Swamp Thing resists embracing this identity, vowing to find a way to become fully human again because "this body imprisons me in more ways than one [...] [and] refuses to help me rid myself of it" (Wein 3: 3).

In the first twenty-four issues, Swamp Thing plays a traditional superhero and encounters robots, dinosaurs, and Batman. He also frequently rescues Abby, the token damsel-in-distress. These adventures take Swamp Thing across the galaxy from the Balkans to New England to outer-space to Gotham and back to the swamps, essentially using the environment as narrative texture rather than featured subject or character. In this way nature is typically used atmospherically to set the horror-vibe of the story. For example, in the opening sequence of the first issue,

> The darkness cries – a long mournful wail that writhes through the gnarled cypress branches like a breath of Hades' wind, skipping over the placid surface of the stagnant mire below. This is bayou country: a swampy desolate marshland forsaken by civilized man – and now inhabited by far less demanding creatures...And this night, this rainy, wind-swept night, impatient humanity intrudes itself into this primitive region once more. (Wein 1: 1)

On a personal level, Swamp Thing resists his own nature and does not see himself as belonging to or serving the environment. Instead, he uses the swamps as a kind of sanctuary, a space where he can conduct experiments to rid himself of his muddy body and regain human form.

Sales for Swamp Thing soon plummeted and the series was cancelled in 1976. With the release of the Wes Craven film *Swamp Thing* in 1982, the character was resurrected under Martin Pasko before being reassigned to the relatively unknown British writer, Alan Moore. Taking over in 1984, Swampy (as he came to be affectionately known as by fans) once again emerged from his muddy lair in a "fusion of the super-hero comic and the horror story" (Campbell 11). In this new telling, Moore made some fundamental changes to the character's origin, most importantly rejecting the creature's human-hybridity and recasting him as a complete vegetative entity. Swamp Thing is depicted as a muscular creature who speaks in halting measured words. In "Loose Ends" and "The Anatomy Lesson," the first two issues published under Moore, he ignores most of the new characters developed by Pasko except for the malevolent Sunderland Corporation, which he refigures as the creature's main foe. Wrapping up the "loose ends" of previous incarnations of the character, Moore opens with Swamp Thing still on a quest to regain his humanity. In the opening sequence he tells the defeated Arcane:

> You were my opposite. I had my humanity…taken away from me. I've been trying to claw it back. You started out human…and threw it all away. You did it deliberately. We defined each other, didn't we? By understanding you…I came closer…to understanding myself. And now…you're dead. Really dead. And what…am I going to do now? (Moore 1: 15)

Following this meditation, the Alec Holland/Swamp Thing origin story is literally shot dead and dissected. Captured by Sunderland, Swamp Thing is placed into a crypto-freezer, allowing Moore to develop an entirely new narrative direction. In the Sunderland labs, the autopsy of the creature is undertaken by Dr. Jason Woodrue. Originally appearing in 1962 as the Floronic Man, an evil botanist with the ability to control botanical life, Woodrue is utilized here to contrast human-plant hybridity with Swamp Thing's emerging all-vegetative identity. As Woodrue performs the autopsy, he discovers pseudo-organs, anatomical structures that resemble human organs, but that do not function in any meaningful way. He realizes that Swamp Thing is not who *he* thinks he is, postulating that when Alec Holland escaped to the bayou he was already dead and "infected" an intelligent plant form with the traces of his consciousness. This creature, with its

> cloudy, confused intelligence, possibly with only the vaguest notion of self, trying to make sense of its new environment…gradually shaping the plant cells that it now inhabits into a shape that it's more comfortable with. It remembers having bones, and so it builds itself a skeleton of wood. It remembers having muscles and constructs muscles from supple plant fiber…it remembers having lungs, and a heart, and a brain…and it does its best to duplicate them…We thought that the Swamp Thing was Alec Holland, somehow transformed into a plant. It wasn't. It was a plant that *thought* it was Alec Holland! A plant that was trying its level best to be Alec Holland. (Moore 1: 48-49)

With this revelation, Woodrue realizes that Swamp Thing is not dead, but simply frozen in a dormant state because "you can't kill a vegetable by shooting it in the head" (Moore 1: 52). When Woodrue warms Swamp Thing's chamber, he awakens to discover his true identity, launching into a profound existential crisis. Unable and unwilling to see that he is not Alec Holland – that he is not and never was human – Swamp Thing kills his Sunderland captors and retreats to the Louisianan swamps.

Lost and confused, Swamp Thing withdraws into nature, losing himself almost to the point of self-annihilation within the 'Green.' The Green is the dimension of elemental interconnectedness that involves all plant life; it is the mind-consciousness or spiritual network of all viridian things. Slowly, Swamp Thing disappears into the swamp, his roots intertwining with his surroundings, his awareness lost in the Green. As Swamp Thing achieves a state of inner peace, he experiences a literal oneness with the earth. While in the Green, Woodrue/Floronic Man, seeking to possess true-plant abilities, finds Swamp Thing and feeds on his body, using the specimens as a medium to make contact with the Green. As he enters this state, he describes,

> I…am…the plant…and beyond the plant? The grass outside…I lie a million silver blades threatening the moon and…and the trees! I…am…the tress. A boa of moss hangs about my shoulders…I feel the intricate genius of the lianas…the giant, timeless wisdom of…the redwoods? But how…far away…is the nearest…redwood? How far am I reaching?...I am withering with a yellow arctic poppy, up the slope of Alaska. So cold….I drift with the seaweed, off Samoa. Somewhere in Russia I incline toward the sun as a field of sighing gold…feel the chrome dustiness of Australia… of Africa…or the Amazon basin. (Moore 1: 82-83)

Because he is not fully vegetative, Woodrue/Floronic Man goes mad from the revelation, shedding him of all traces of humanness. He emerges as an eco-terrorist obsessed with protecting nature from humans, a belief that metastasizes into a form of cancerous biophilia. Hailing the coming of a "green revolution," Floronic Man attempts to destroy all non-plant life, heralding

> the green millennium...I am one with the wilderness, it's will works through me...For I asked of it, saying what would you have me do? And it said, 'purify' and it said 'destroy' 'destroy all the creatures that would destroy us, that would destroy the ecosphere with their poisons and bulldozers, cut them down like blighted wood. Let us have another green world!' (Moore 1: 107)

As Floronic Man begins to destroy people in the name of trees, Swamp Thing is aroused from his slumber by Abby. Recognizing for the first time the extent of his physical power, Swamp Thing thwarts Floronic Man by explaining that plants and humans are co-dependent – without one there is no other because "what will change the oxygen…back into…the gasses that…we need to survive…when the men…and animals…are dead?" (124). This episode, the climax of the story arc developed in issues #20-27, injects the narrative with ecological awareness. The surface conflict with Floronic Man reveals Moore's criticism of humanity's anthropocentrism and eco-phobic stance when interacting with the natural world. Additionally, Moore challenges how readers interact with the environment, forcing them to directly engage with nature, communing with the terrifying and terrified Swamp Thing without any human medium to cushion the experience. In this way, nature is pushed to the front-and-center of the Swamp Thing narrative; it becomes its lead character, main subject, and narrative axis.

In issues #28- 34, Swamp Thing embarks on a quest to discover his true identity. In the opening sequence, he revisits Alec Holland's story by finding and interring his physical remains, finally putting the past – the one he thought was his own – to rest. In revisiting the character's origin story, Arcane, the original Swamp Thing nemesis, also makes a reappearance. After banishing Abby's soul to the underworld, Arcane thinks he has comes to wage his final battle with Alec Holland, to which Swamp Thing responds "this…is our…first battle…you…have never…encountered *me* before" (Moore 2: 99). With this statement,

Swamp Thing extricates the last of Alec Holland from his self-construct. Liberated, he assumes his place in nature, declaring,

> This…is our…*final*…*battle*…you claim…the earth…for your own…Arcane…yet you have not…one *tenth*…of the Earth's power. You…mock…*God!* You make…the world…*worthless*…with your lies! How…dare…you?? (Moore 2: 100).

After Arcane is defeated and banished to Hell, Swamp Thing embarks on a quest to recover Abby's soul. Entering the Green, Swamp Thing is guided through the lands of the "just dead," Paradise, and Hell. When he seeks permission of Spectre, "the sentinel of [the] dismal territory," (Moore 2: 128) to enter hell, Spectre reveals to Swamp Thing that he is an earth elemental – an entity long thought to be extinct. Unable to question Spectre further, Swamp Thing is guided by Etrigan, a couplet-rhyming demon, through the Dantesque underworld where he encounters both Sunderland and Arcane and barely escapes with Abby.

Reminiscent of classic heroic quests, Swamp Thing's journey to find Abby ultimately leads him to discover his own identity and purpose. After escaping hell, Abby experiences a dream in which she revisits the original Alec Olsen/Len Wein/Swamp Thing story. There it is revealed to her by the biblical Abel that Alec Holland "was not the first thing to walk the swamps! There were others before him […] sour times are returning to your world, and your world has once again shaped a *protector* to stand *against* them" (Moore 2: 190). Though she forgets the dream the moment she awakens, Abby still manages to help Swamp Thing discover his true potential. In the next scene, the two make love for the first time, consummating their union by her ingestion of psychotropic tubers that grow within his body. The panels of the subsequent pages are saturated with glistening creatures and plant-life, the ultimate communion of plant and animal:

> Together we bathe in raw life: Honey rolls across our tongue. The fragrance of decay, mesmeric and overwhelming, excited our nostrils. We savor both equally…for life is not all that we comprehend…There is no contradiction…Only the pulse. The pulse within the world. Within us. (Moore 2: 217)

In this joining, the two experience the full breadth, depth, and complexity of their interdependence, recognizing that plant and animal are fused within the very heartbeat of the natural world.

In issues #35-42, Moore begins his American gothic storyline in which he addresses sexism, racism, and environmentalism while mining a vast tradition of horror literature. Led by mysterious magician John Constantine, Swamp Thing is sent on a kind of scavenger hunt across the United States, each tale a clue in discovering his true identity and purpose. One of the most interesting characters Swamp Thing encounters is Nukeface, a radioactive tramp addicted to nuclear waste. Homeless, lethal, and incredibly patriotic, this man lectures all his victims on the importance of nuclear energy to American industry bemoaning "all that hippy anti-nuclear stuff" (Moore 3: 19) that impedes progress. When nuclear waste is dumped illegally in the swamps, the man unknowingly contaminates and destroys Swamp Thing's verdant body. As he decays, Swamp Thing discovers that he is recyclable:

> I have...regenerated...using plant tissue...untouched...by the bio...restorative formula...I do not...need it...anymore...I am learning...so much...about what I am...and the knowledge...is gradually...changing me...But what...is it changing...me into? What...am I...becoming? I bide my time...in this place...until I grown...and I consider...this organism...that I am. Sometimes...I am in awe...at its strangeness...and complexity...sometimes... I am almost frightened...by my own possibilities. (Moore 3: 62)

Swamp Thing regenerates in a pure form, emerging stronger in a body unaffected by the human-engineered bio-restorative formula. When Abby later sprays him with insecticide he complains that it irritates his still-forming body. In this interplay between science and nature, Moore explores the relationship between industrial progress and the environment. While Swamp Thing is able to regenerate, the reader is told of others, both people and places, who have been irrevocably damaged in the name of progress. This information is provided through a scrapbooking technique in which the panels of the text are layered with newspaper clippings. These snippets cover a variety of environmental-themed topics including toxic waste, acid rain, and dioxin testing. One page – pushing beyond the panel lines as if to symbolize growing prevalence or massiveness – includes the testimonies of people impacted by 'progress,' suggesting the destructiveness of industrial growth and its depletion of the restorative powers of nature.

This episode seems as though it were a response to Rachel Carson's *Silent Spring.* Published in 1962, *Silent Spring* is largely credited for inspiring the modern environmental movement in the United States. In her work, Carson contests the widespread use of pesticides and provides evidence indicating its destructiveness on the environment. Her work was one of the first to sound the alarm and demonstrate through research the damaging impact of industrial progress on nature. Moore seems to be echoing these concerns in Swamp Thing's interactions with Nukeface, questioning what is gained through unreflective progress and why environmental concerns or critiques of industrial practices are often deemed un-American.

This tension between progress and nature is revisited throughout these issues, and Swamp Thing is cast as the earth's protector and restorer. After ridding the swamps of nuclear waste, Swamp Thing is sent to destroy a colony of underwater vampires. Initially he is overpowered by them because, as he tells himself, "I am …too human…in the way I think…in the way I fight…I must learn…to exploit…the possibilities of what I am" (Moore 3: 109). Rooting himself in his identity, Swamp Thing harnesses the power of all living things:

> I feel my way…into the roots that knot…within its powdery soil…I gather myself…in the heart…of the rootweb…its tendril…become my sinews…my arms…are two miles long…encircling…the stale obsidian depths…I begin…to flex…my muscles." (Moore 3: 113)

Here, Swamp Thing appears in the center of the page like Da Vinci's Vitruvian Man, radiating outward, encompassing the entire scene. With this energy, he shrugs his "vast buried shoulders" (Moore 3: 114), the foundations of the earth, and drains the bayou, destroying the vampires by exposing them to the bright sun. As they are vanquished, the vampires protest the destruction of their civilization and their entitlement to social progress. Similarly, in a later episode, Swamp Thing is unable to assist a female werewolf who encompasses the rage of subjugated and sexualized womanhood as if to comment on the 'natural' place of women within a patriarchal society. Along the same lines, in the final story, Swamp Thing helps a plantation of zombies redress their racist past and literally put old "ghosts" to rest. These episodes cast Swamp Thing as maintainer of holistic earth, an assistive force within a Gaia-framed natural world. This characterization evokes the Gaia Hypothesis, a theory developed by scientists James Lovelock and Lynn Margulis in the 1970s, which proposes that the entire earth is a self-regulating entity in which all life-forms and nonorganic life interact with one another to maintain a vital biosphere. In this way, Swamp Thing empowers living things to restore balance within a socially unjust and industry-driven world.

In issues #46- 50, Constantine reveals to Swamp Thing that a powerful evil is coming that will threaten all existence. As he leads Swamp Thing through the jungles of Brazil, he reveals to him the Parliament of Trees and asks him to appeal for their support in the coming battle. Located along the river Tefé and inaccessible to humans, the Parliament is a resting place for previous caretakers, earlier swamp things. There the reader encounters Alec Holland and learns of many other previous swamp things including Alex Olsen. When Swamp Thing asks the Parliament for help, he is told,

> Power is not a thing. To be calm *within* oneself, that is the way of the wood. Power tempts anger and is like wildfire…*avoid* it….flesh doubts. Wood *knows.* If you wish to understand *evil,* you must understand the *bark,* the *roots,* the worms of the earth. *That* is the wisdom of an *erl-king. Aphid* eats *leaf. Ladybug* eats aphid. Soil absorbs dead ladybug. Plant feeds upon *soil.* Is *aphid* evil? Is *ladybug* evil? Is *soil* evil? *Where is evil in the wood*? (Moore 4: 113)

Swamp Thing rejects the Parliament's position, arguing that nature must involve itself with the affairs of the world. Unsupported, but determined, Swamp Thing assumes a central role in the cosmic fight against evil.

The foreboding doom is soon revealed as the original Darkness, the "the soul of darkness itself…a complete absence of divine light" (Moore 4: 166). Released and overwhelming, no one is able to subdue the Darkness because none can explain to it its creational purpose. As the blackness invades heaven, Swamp Thing alone volunteers an explanation,

> I have seen evil...its cruelty...the randomness with which it ravages innocent and guilty alike... I have not understood it. I asked the parliament of trees whose

> knowledge is older, greater than mine...they seemed to insist that there was no evil...but I... have seen evil... and their answer was incomprehensible to me and yet... and yet... they spoke of aphids eating leaves bugs eating aphids themselves finally devoured by the soil... feeding the foliage... they asked where evil dwelled within this cycle and told me... to look to the soil... the black soil is rich in foul decay... yet glorious life... springs from it but however dazzling the flourishes of life in the end... all decays to the same black hummus. Perhaps evil is the hummus formed by virtue's decay and perhaps it is from the dark sinister loam that virtue grows strongest? (Moore 4: 194-195)

Of all the creatures of heaven and earth, Swamp Thing alone satisfies the Darkness by explaining that good and evil are interdependent and must coexist. Like plant and animal, industry and nature, Darkness and Light must join hands and suffuse the earth with their synchronicity. Disaster is averted, but, fundamentally altered by the union of good and evil, of dark and light; Swamp Thing seems to achieve ecological vision.

This awareness is soon challenged, as issues #51- 56 deal with Abby's arrest for having inappropriate relations with a nonhuman, Swamp Thing. Reacting to her imprisonment with ferocious anger, Swamp Thing invades Gotham City (where she is being tried) and overwhelms it with a riot of vegetation:

> The city is all about him, a defiant surge of stone and steel and glass that forces back the surrounding wilderness, jealously establishing its rigid territory. The swamp god flexes his mind. The wilderness shrugs. All over town from sudden cracks and fissures, the sidewalks begin to bleed emerald. Moss dribbles up the sheer sides of glass towers and the ghettos are burning with orchids…rustling through the foliage, the swamp god shivers with satisfaction. Why? Why had he bothered to restrain his glorious power for so long? Why hold back paradise? (Moore 5: 53-54)

As he turns the asphalt kingdom into an exotic jungle, Swamp Thing not only displays the versatility and bountifulness of growing things, but demonstrates the puniness of man in comparison to the omnipresence of nature. Although he is initially feared by Gothamites, he is soon enshrined and worshipped as a god. Realizing the unsustainability of the situation, Batman eventually negotiates for Abby's release. As the switch-off is happening, Swamp Thing is destroyed through an electromagnetic form of napalm.[7] His bio-electric frequency thusly scrambled, Swamp Thing's consciousness spirals into deep space, where he thinks of his relationship with Abby, deciding that "the distance between [them] is unbridgeable" (Moore 5: 140). Destroyed and unable to return to earth, Swamp Thing believes that humans and nature can never live in harmony, not without one overpowering the other.

Unable to return to earth, Swamp Thing floats to distant planets and outposts in the galaxies. For the remaining issues in the series, the narrative switches genres from southern gothic horror to a fantasy sci-fi. Initially, Swamp Thing attempts to create a replica of Abby and their home. He quickly destroys his creations, angrily realizing their futility and inauthenticity, suggesting that the two are inexplicably and unhappily bonded: "oh, Great Mother *Earth*...who…am I *talking* to..? What *madness*...is upon me…?" (Moore 5: 161).

As he floats through the cosmos, Swamp Thing engages with a variety of other life forms. On one planet, he helps rebuild an ecosystem and save a famine-starved populace. At some point, in deep space, he unwillingly mates with a digital cosmic entity. As the entity relays the story to her offspring,[8] the reader learns of this communion, this fusion of machine and nature:

> He fell past the depthless secrets of my entire race and their procreation, to him an unreadable manuscript illuminated by fabulous and baffling illustrations. Your transparent pods hung trembling from my iron stems among the spiny husks of older and unfertilized eggs…some of the pods had rapidly acquired their fur of circuitry, twittering drones already floating towards them with cheeks puffed out and mouths full of genius. In the hothouse of my womb, children were blossoming. You were blossoming. (Moore 6: 100)

After Swamp Thing frees himself of the entity, she reflects,

> If he could have lingered, if he could have looked beyond those chip-encrusted pod walls and seen you as I see you now…would he see any facet of himself reflected? The darkness of your eyes, perhaps, so like his own?...Could he have

[7] To emphasize this theme, one of the onlookers appears with a pin reading "No More Chernobyls."

[8] It is interesting to note that the lettering also changes here and appears typed rather than handwritten.

> loved these lukewarm and peculiar little creatures; these fetal bio-automata; you, his children." (Moore 6: 102)

Here, Moore seems to be considering the uncertain future in which nature will be forced to contend and even 'mate' with a digitized world.

Eventually, Swamp Thing makes his way to a sentient vegetative planet where they teach him to control and restore his electromagnetic frequency. Making his way back to earth, Swamp Thing learns that his place, his destiny, is in the center of everything as a steward of the entire planet. By assisting a celestial named Metron, he experiences "a multitude of universes collapsing upon each other, attempting unification and a single coherent cosmology" (Moore 6: 145). Renewed with a sense of holistic purpose, Swamp Thing returns to Abby to discover that she has arrived at a similar position on the relationship of man and nature,

> sometimes I think for us to really help the environment, we'd need a different world- someplace that taught people how everything from the environment to politics was connected, taught them to think and take responsibility...I see people working on that but it never gets off the ground. (Moore 6: 191)

Swamp Thing encourages her not to be dispirited because "perhaps a new *world*…is closer than you *think*…ready to soar aloft…from the ashes of the *old*" (Moore 6: 191). Though he hopes for a better future, Swamp Thing decides not to actively create it because

> If he should feed the world…heal all the wounds man's smoldering industries have made…what would he do? Would he renounce…the wealth his sawmills bring…step gently on the flowers instead…and pluck each apple with respect…for this abundant world…no. He would pump more poisons…build more mines…safe in the knowledge that I stood on hand…to mend the biosphere…endlessly covering the scars he could now endlessly inflict. (Moore 6: 196)

Finally understanding the stance of the Parliament of Trees, Swamp Thing resolves that humankind must be responsible for its interactions with nature. Retreating, he creates a home for himself and Abby within the swamps and the two retire indefinitely.

Moore's Eco-legacy in Comics

Though the series is still active, under Moore's tutelage, Swamp Thing's monthly readership increased five-fold (Parkin 37). It also won many of the industry's highest honors, including several Jack Kirby Awards. Moore's Swamp Thing is multilayered and thick with meaning, and it was one of the first graphic works to present a complex and intelligent view of the environment. Rejecting the naïve and dichotomous worldview of earlier comics, Moore's material conveyed an affective earth, a *feeling* and personable space with which people must not only coexist, but be responsible to. In this way, Swamp Thing's complicating portrait of the relationship between plant and animal heralded an abiding sense of eco-consciousness within the world of comics, altering the treatment of nature as both character and subject.

Following Moore's example, a new cast of eco-villains and heroes emerged in the mid-1980s. Poison Ivy and Black Orchid, two existing characters, were given new origins, recast as former colleagues of Alec Holland who together studied under Dr. Jason Woodrue.[9] Poison Ivy, once a seductress figure, is now an eco-terrorist and one of Batman's most vexing villains. In one story, she entraps all of Gotham City's corrupters in a man-eating plant. Tefé Holland, Abby and Swamp Thing's daughter, is a human-plant hybrid and unsure of whether to use her formidable powers on behalf of plants or animals. These characters present a multidimensional view of man's continuous struggle to foster a sustainable, responsible and harmonious relationship with nature.

---

9 In this way, all human-plant hybrid creatures share a single origin within the DC universe.

Works Cited

Austin, Richard C. *Baptized Into Wilderness: A Christian Perspective on John Muir.* Abingdon, VA: Creekside Press, 1991. Print.

Campbell, Ramsey. Foreword. *Saga of the Swamp Thing Vol. 1.* By Alan Moore. New York: DC Comics, 2012. 9-11. Print.

Coogan, Peter. “The Definition of the Superhero.” *A Comics Studies Reader.* Eds. Jeet Heer and Kent Worcester. Mississippi: University Press of Mississippi, 2009. 77-93. Print.

Daniels, Les. *Superman: The Complete History: The Life and Times of the Man of Steel.* San Francisco: Chronicle Books, 1998. Print.

Ferber, Michael. *A Dictionary of Literary Symbols.* New York: Cambridge University Press, 1999. Print.

Homer. *The Iliad.* Trans. Robert Fagles. New York: Penguin Books, 1998.

Moore, Allen, writer. *Sage of the Swamp Thing. Vol. 1.* 1983-84. Reprint. Inks by John Totleben. Pencils by Stephen Bissette. New York: DC Comics, 2012. Print.

—. *Saga of the Swamp Thing, Vol. 2.* 1984-1985. Reprint. Inks by John Totleben. Pencils by Stephen Bissette. New York: DC Comics, 2012. Print.

—. *Saga of the Swamp Thing, Vol. 3.* 1985. Reprint. Inks by John Totelben. Pencils by Stephen Bissette. New York: DC Comics, 2012. Print.

—. *Saga of the Swamp Thing, Vol. 4.* 1985-1986. Reprint. Inks by John Totelben. Pencils by Stephen Bissette. New York: DC Comics, 2012. Print.

—. *Saga of the Swamp Thing, Vol. 5.* 1986-1987. Reprint. Inks by Rick Veitch. Pencils by Alfredo Alcala. New York: DC Comics, 2012. Print.

—. *Saga of the Swamp Thing, Vol. 6.* 1987. Reprint. Inks by John Totelben. Pencils by Rick Veitch. New York: DC Comics, 2012. Print.

Oates, David. “Some Questions about Sexless Nature Writing.” *Postwestern Cultures: Literature, Theory, Space.* Ed. Susan Kollin. Nebraska: University of Nebraska, 2007. 115-126. Print.

Parkin, Lance. *Alan Moore...*United Kingdom: Pocket Essentials, 2001. Print.

Pevey, Timothy Aaron. "From Superman to Superbland: The Man of Steel's Popular Decline among Postmodern Youth." Diss. GSU, 2007. Print.

United States. Cong. Senate. Committee on the Judiciary. *Comic Books and Juvenile Delinquency.* Washington: GPO, 1955. Print.

Wein, Len, writer. *House of Secrets #92.* Art by Bernie Wrightson. New York: DC Comics, 1971. Print.

Weiner, Stephen. *Faster Than a Speeding Bullet: The Rise of the Graphic Novel.* New York: NBM Publishing, 2003. Print.

Weisinger, Mort. "Aquaman: The Submarine Strikes." Art by Paul Norris. *More Fun Comics #73.* Ed. Mort Weisinger. New York: DC, 1941. Print.

Worster, Donald. *A Passion for Nature: The Life of John Muir.* New York: OUP, 2008. Print.

Reproducing Plant Bodies on the Great Plains

Aubrey Streit Krug

Abstract: What might be gained from thinking through plants and humans as growing, reproducing "bodies"? How do plant bodies produce human bodies, and vice versa, on the Great Plains? This essay explores these questions by analyzing how cultural narratives of power are reproduced in public agricultural rhetoric about plants as well as in the material practices of propagating and eradicating plants. The conclusion reflects on the role critical plant studies might play in academic and popular cultures by posing a new question: What ways might a non-normative or "queer" way of thinking about plant reproduction help us solve on the Great Plains?

> *Georgic may be understood not merely as a genre but more importantly as discursive and material practices fundamentally concerned with articulating relations between 'man' and nature and between work and leisure, the rights of property, the founding of nations and empires, and cultivation or farming as a metaphor for civilization. Moreover, georgic discursive and material practices conventionally inscribe the sexual division of labor, patriarchal lineage, and heterosexual reproduction as foundational and immutable, as the order of nature.*
>
> Jill H. Casid

An example of the sort of "georgic discursive and material practices" discussed by Jill Casid can be found in a 2010 television commercial for Dow AgroSciences' herbicide "PowerFlex." Here, a weed killer is depicted as an armed battleship sailing through a winter wheat field to a lively, patriotic tune. A caption on YouTube explains, "PowerFlex herbicide is the flagship of reliable performance in winter wheat. It controls tough, yield-robbing broadleaf weeds and grasses. Watch this video to see how PowerFlex helps growers take command of their fields" (Bader Rutter). The commercial presents a militaristic clash. The agricultural field of waving grasses becomes a battlefield of waves at sea, and the enemy is vegetation that needs to be controlled or eliminated. In the historical context of settler colonialism on the frontier of the North American Great Plains, this threatening plant life can be linked to the relationship between indigenous and immigrant human cultures. During the nineteenth-century settling of the frontier, vegetal communities of the prairie were plowed up, animal communities of bison were hunted down, and human communities of Native Americans were killed, forced onto reserves, and threatened with assimilation. The PowerFlex commercial continues to portray this settler colonial perspective. Adaptive invasive and native "weeds and grasses" labeled "tough" are deemed bad because they are "yield-robbing" while the cultivated "winter wheat" is implied to be good because it is a cash crop; it is right to use and own the land to make money. To support the American mythos of founding civilization in the wilderness, the wild weeds and grasses are an invisible threat never actually shown in the commercial, but they are assumed to have agency (the farmers need an herbicide to "take command of their fields" away from them). However, the cultivated crop is visually portrayed as passive, docile, and fully domesticated. The ground appears reassuringly settled. The land is properly used, owned, and monitored, and the colonizing rhetoric of controlling "nature" is emphasized.

What may be less obvious, however, are the ways in which this rhetorical and material example of the georgic "inscribe[s] the sexual division of labor, patriarchal lineage, and heterosexual reproduction as foundational and immutable, as the order of nature" (Casid xxii). After all, there are no human bodies visually represented; only the technology of the battleship and the swaying bodies of plants are shown against the earth and sky. Yet consider how the commercial, which features an apparently masculine narrator with a deep voice, might speak to anxieties about gendered control over the field, nation, or family by visually portraying a singular, sleek battleship (housing projectile weapons and surveillance devices) penetrating and patrolling the cultivated landscape of winter wheat. In this way, the commercial connects to the settler colonial rhetoric promoted by the United States and Canadian governments, which historically have appealed to masculine authority and patriarchal marriage institutions for the protection of women, children, and heteronormative families back home as well as on the frontier; these appeals sanctioned violent and militaristic actions against natives, against non-monogamous or non-patriarchal relations, and against wilderness in the West (Carter). Notice also how the narrator addresses human audiences as "growers," a term that might be used to refer to plants as well. Based upon this, I would suggest that plants and humans on the Great Plains are inextricably linked through a narrative of reproducing plant bodies. The underlying claim of the advertisement is that human "growers" must control or be controlled by plant "growers," and that the power of control can be flexed by the masculine human body (individual as well as social) in a two-fold process of agriculture that is mediated by human technology: stopping non-desirable plant bodies from reproducing in order to purposefully reproduce other, more desirable plant bodies.

In this essay, the site of reproducing plant bodies on the Great Plains is a place for agricultural and intellectual inquiry. First, I seek a better understanding of the reproductive relationships between humans and plants. I ask: What might be gained from thinking through plants and humans as growing, reproducing "bodies"? How do plant bodies produce human bodies, and vice versa, on the Great Plains? Then, I examine three noteworthy contemporary texts in public circulation: another commercial for herbicide targeted at farmers, an online succession planning guide for family farms, and a textbook on plant reproduction used by students in horticulture. Through studying these texts, I explore how cultural narratives of power are reproduced in public agricultural rhetoric about plants as well as in the material practices of propagating and eradicating various plants. Finally, I consider the roles critical plant studies might play in academic and popular cultures. In particular, I wonder: what problems might a non-normative or "queer" way of thinking about plant reproduction help us solve on the Great Plains in particular, as well as in other agricultural communities more broadly?

****

When we think of plants, we tend to think of growth. Plants in human discourse are often thematically associated with narratives of development. To plant a seed, literally or figuratively, means to desire the sprouting and rooting of that seed into a vegetal form that bears material or immaterial harvest. To plant a stake or a flag means to lay claim to physical ground or to a concept, to own or possess it and thus provide for the growth of one's property, community, or nation. To plant a kiss might signal something similar, the growth of a bond or relationship between people. In each of these images of plants and growth there is the theme of controlled personal or cultural reproduction with regard to the body: the act of planting either sustains the individual human body or a connection between human bodies, such as a larger human socio-political body. In each of these images of reproduction, the body – at whatever scale – is formulated in terms of desire, possession, intimacy. Planting something, or biologically or culturally reproducing some kind of plant(ed) body, becomes something more: a sexual act that takes on certain forms and calls for the construction of certain norms and genders with regard to human bodies.

Furthermore, as plants are gathered, classified, sown, eaten, and bred, plant bodies become linked to the life and health of human bodies. Simple distinctions between plants and humans and other animals begin to break down when perceived through multiple conceptions of the body, from individual to cultural, from political to material. Through the acts of harvesting and eating, plant bodies form the beams, walls, floors, and siding in the structures of animal bodies; plant bodies are what allow for the growth of human bodies in the form of populations. Pulped and pressed plant bodies have become the pages of books. As humans have continued to develop their technology, fossil plant and animal bodies that captured ancient sunlight have increasingly been used to fuel the movements and economies of contemporary human bodies. Plant bodies, then, are in part what allow for the growth of human cultural and political bodies. Cultivating, consuming, and constructing human bodies with plant bodies, or using plant bodies to reproduce human bodies, becomes something more: an ecological act that takes on certain forms and calls for the production of certain locations with regard to human bodies.

The reproduction of plant bodies, then, is both sexual and ecological, and is one place where gender theory and ecocritical theory meet. It holds particular potential with regard to queer ecology, a field of analysis based on the idea that "there is an ongoing relationship between sex and nature that exists institutionally, discursively, scientifically, spatially, politically, poetically, and ethically" (Mortimer-Sandilands and Erickson 5). But this site is also something more than sexual and ecological: it is economic. In each of the above explorations of plants, plant bodies become commodities, tied to the reproduction of capital in the present for the future. For instance, the cultivation of certain plant bodies for human consumption on a piece of land is what has been culturally and legally deemed "proper use" of the land, based on a rationale of economic production.

We define producers scientifically as plant bodies that capture solar energy (such as primary producers in a food web) as well as culturally as human bodies (such as farmers) that produce plant bodies. Production is conceptualized as an ongoing series or succession, in which a producer might reproduce itself, something similar to itself, or something unlike itself. Furthermore, produce as a noun is often used to represent the outcome of a "natural" process of production: think of the produce aisle in a supermarket. This assumption of "nature" is interesting since there is no similar assumption of naturalness in a human cultural construct like an idea, or a piece of construction like a human built environment (and it is the metaphor of construction that is most often used in feminist, gender, and queer theories). Therefore, the energy that goes into the production of bodies, as well as into their cultivated and controlled reproduction, is discursively naturalized in a way that calls for our attention.

Stacy Alaimo's concept of trans-corporeality questions this naturalization and dematerialization of the body (2). Trans-corporeality helps us understand how the boundaries of the human body are constructed as well as crossed, and how these trans-movements connect and constitute the human body with the more-than-human world through an ongoing process of mutual (re)production. Trans-corporeality explains how what we think of as "human" is produced in the context of non-human creatures and non-living entities, and by flows of matter across borders of self, safety, and skin: think not only of plants and animals eaten as food, but of fungi in soil, bacteria in guts, and chemicals in air and water. The human body, like the plant body, can be both producer and produced.

A sense of plant bodies as materially discursive might help us understand why certain narratives about plants continue to be reproduced in human discourse and rhetoric, and how repeated patterns and acts reproduce sexual, ecological, and economic understandings of plants with regard to growth. Since rhetoric can be defined as language used for a purpose – language constructed and used in such a way that it *does something*, or attempts to do something, in the context of the human audience and/or more-than-human world – rhetoric itself can be seen as powerfully productive. If rhetoric is the art of discourse, rhetoric works to produce and reproduce power, inscribing and reinscribing itself rather than offering a non-material, subversive escape route. A rhetorical reading of plant bodies, then, might lead us to examine not how plant bodies escape discipline and power but rather, first, how plant bodies are rhetorically, materially situated within discipline and power.[1]

Finally, plant bodies have a particular relationship to place. My inquiry is focused on the Great Plains, a physiogeographic province that is a material but also imagined place – a produced body with constructed boundaries, if you will. The Great Plains contains various prairie bioregions that are defined, at least in part, by their vegetation. Humans use the composition and classification of plant matter with regard to time and space to construct the concept of a bioregion. The tallgrass prairie, for instance, is known in terms of vegetation/climate/topography that forms a place full of grass, crops, memories, words, dreams, losses: in other words, it includes certain plant bodies and human animal bodies, as well as being a kind of body in itself. The body thus becomes central to the question of location, context, and rhetorical situation (where are we?) and to the question of identity, boundary, and forms of life (what are we?) as well as to even more questions about economics, ethics, and desire (what are we to do here?). If the plant bodies that human bodies contain, cultivate, consume, and contribute to all produce identity, then to cross bodies, connect them, critique them, and live them – all of these ways of being and doing – necessarily raise questions about the reproduction of a place.

From a social perspective, human reproductive rates on the Great Plains include both diminishing and aging white populations in rural areas and small towns due to outmigration to cities and urban centers in the Midwest and on the coasts, and increasing and young non-white populations of in-migrating Chicano/a cultures and on Native American reservations. From an ecological perspective, plant reproductive rates also vary. They include the plowing up of native tallgrass, mixed grass, and short grass prairie plant communities, along with efforts to

[1] This is not to suggest that there is no place for subversion, but instead that bodies cannot be subversive simply by trying to combat production. The way to subvert might be to (counter-intuitively) produce something different, to produce opportunities and directions for movement – movement of the kind that, like resistance for Michel Foucault, doesn't liberate bodies from power, but does *precede* power, always a step ahead of it, with discipline a step behind.

conserve and restore these polycultural ecosystems; the astonishing growth of Ogallala-aquifer and dammed-river fed fields of annual monocultural crops, especially corn; and the efforts to gather, breed and seed certain plant bodies, be they from native seedbanks or patented germplasms. Building upon this, and upon what I see as the central claim of the PowerFlex herbicide commercial discussed earlier, I would argue that the struggle to reproduce more desirable kinds of *plant* material bodies (at the expense of others that threaten to reproduce) is rhetorically represented as necessary for the reproduction of more desirable kinds of *human* social bodies (at the expense of others that threaten to reproduce) on the Great Plains. This is one narrative of reproducing plant bodies, but as I will suggest later, it need not be the only one.

****

I saw the Capreno Herbicide commercial "Man vs Weed" for the first time on television in Nebraska in spring 2013, though it was published on YouTube in May 2012 (BayerCropScienceUS). The 30-second spot follows a familiar storyline for anyone who has seen such cable television shows as the Discovery Channel's *Man vs Wild* (in which the host is dropped into "wild" locations and then has to figure out ways to shelter and provide for himself and return to "civilization") or the Travel Channel's *Man vs Food* (in which the host travels to restaurants around America to take on "food challenges" based on portion size or ingredients). These television shows dramatize the masculine human host's bodily survival or health. Though audiences know the answer to the question of "Will he make it through this battle alive?" we watch anyway for reassurance that "nature," whether it comes in the terrifying form of climatic extremes or of a plate-sized cheeseburger, can be overcome through "man's" mental ingenuity or sheer willpower.

The threatening "nature" in the Capreno Herbicide commercial are weeds, plants out of place that might reproduce themselves and upset the carefully cultivated crop that "naturally" belongs in the field. Even though weeds are as "natural" or "normal" as any other plant in their growth patterns, adaptive abilities, and reproductive tendencies, they are implied to be "unnatural" because they are reproducing in the wrong place and undermining the "normal" profitability of that place. Furthermore, these non-normative weeds are given some kind of agency in their ability to threaten the well being of the human species. As in the PowerFlex herbicide commercial, the weeds are not visually depicted as growing or proliferating; instead, it is the voice of the narrator who gives viewers the sense that these wild, fertile weeds will overtake the domesticated place of agriculture if they are not sterilized or exterminated. The masculine farmer-host later says enigmatically, "Weeds this year – toughest ever." The weeds must be personified into a worthy enemy in order to justify the chemical warfare enacted against them. Later, viewers see shots of singular weeds or small groups of them, which stand green and innocuous until they wither under the application of herbicide.

In the commercial, the weeds are not given particular species names. Instead they are grouped by general categories: grasses, broadleaf weeds, and glyphosate-resistant weeds. The inclusion of this last group of weeds foregrounds what is rhetorically unremarked upon: namely, that some weeds have become a reproductive threat not in spite of but rather *because of* the application of manufactured chemical herbicides like glyphosate, also known by its twentieth-century commercial name of Roundup. The adaptive group of glyphosate-resistant weeds was not created by human actions, but it was encouraged to reproduce by them. Since the application of glyphosate to fields destroys their competitors (i.e. non-glyphosate-resistant weeds), the resistant weeds are selected for with each growing cycle. When farmers and agricultural experts refer to resistant weed varieties as "superweeds" in the media and in popular culture, they are therefore subtly calling attention to the ways these weed populations have been produced. Superweeds are also often seen as vegetal bodies that – in the word of the Wikipedia entry on "Glyphosate" – have "afflicted" agricultural land, which makes them sound unhealthy and dangerous to the social body of humans ("Glyphosate"). The adaptive abilities of weeds are well documented and known to corporations like Monsanto, which also clearly defines "herbicide selection pressure" ("How Does Weed Resistance Develop?"). But these corporations, and the commercials they produce, have an interest in persuading farmers to believe that with the help of (corporate) technology, humans can stay one step ahead of the weeds that through their actions they encourage to reproduce. Most farmers of crops like corn are bound by similar economic circumstances that compel them to agree; without government subsidies, seedstock from corporations, and outside chemical inputs (fertilizers, herbicides, and fossil-fuel-run machinery), they would not be able to pay the mortgage or rent on their land and would be displaced from it. The current American agricultural policy system, supported by legal practices and corporate lobbying and reinforced by a cultural value system that uncritically re-deploys settler colonial logic, affirms its own rhetorical and material reproduction by acting as if there are no alternatives.

Returning to the commercial, the battle between human bodies and plant bodies is depicted as an annual event, with references to "This season on *Man vs Weed*" and "Weeds this year." Our host is Ethan, a tanned white man, who drives a large pick-up truck down a dirt road and stoically puts on his cap, as if uniforming himself for combat. Ethan's secret weapon is this season's plot development: he "discovers" Capreno Herbicide. This discovery is not described in terms of cost but rather glossed as a natural development, implying that the masculine Ethan is intelligent and resourceful because he happened upon this solution. The announcer adds that Capreno has the "longest residual of any post" – technical language which, to the informed viewer, means that of the herbicides you would apply to corn after it has emerged from the ground, Capreno will stay in the soil the longest. Like Ethan, whose discovery of how to enhance the extraction of profit from his corn crop will allow him and his crops to remain settled on the land he rents or owns, Capreno herbicide is granted credibility because of its temporal stability, its ability to persist through time in a particular place (as long as rain doesn't fall within an hour of its application to a field, it even has a guarantee of "rainfastness," according to the "FAQs"). Capreno thus becomes not an agent of death, but of the preservation of life. Through materialist rhetoric, the advertisement establishes the herbicide, like the crops in the fields, as a way to reproduce plant bodies that are natural and normal.

The commercial's imagery shows weeds located between orderly rows of homogenous corn plants, literally in between the planted lines of legibility, and though no machinery or implements are portrayed (we do not see Ethan wielding any of the technological tools of battle or interacting with the chemicals he applies) the animated weeds brown and shrink back under the mere gaze of the viewer in the commercial. Viewers are given an intimate, ground-level perspective on the field. But as soon as the weeds are dead and the season's productive mission is accomplished, the camera rapidly pulls directly up and out so that viewers see a restoration of the orderly rows of crops, then the single corn field, then the broader section of land with fields and roads visible through a few gauzy white clouds. This is the familiar landscape of an aerial survey that shows us again how the land has been conquered; in James C. Scott's evocative phrase, viewers are "seeing like a state" as they take in the perspective of the cadastral survey used to map and tame the land during the settler colonization of the United States. The announcer asserts the logic of colonization in the American West: "Capreno. Because out here, it's man versus weed." And the camera returns to Ethan, who chimes in, "And man will emerge . . . victorious." Capreno, Ethan, and the collective noun "man" all stand for human, reassuring audiences of the power of raced-white, gendered-masculine authority to use technology in order to agriculturally control and normalize nature for the good of the nation.

As the herbicide logo flashes back onto the screen, superimposed over the image of Ethan standing in a field of corn that has now grown almost to his shoulders, the closing line reminds viewers that Capreno offers an "amazing end-of-season clean." The herbicide purifies or cleanses the ironically dirty or messy agricultural field from the weeds that afflict it, ensuring the safety and health of the properly reproducing plant bodies—and the safety and health of the economies of human bodies that the reproduction of these plant bodies supports. For now, at least.

What will happen to this farm as the human body of Ethan, a "top producer" in the language of agribusiness publications[2], ages? What will happen if Ethan is injured in a mechanical accident or becomes ill due to prolonged or haphazard chemical exposure? This is where "succession planning" comes in. A "succession" most commonly describes the legal, political passing down of power, in the form of a title or a piece of property, through an orderly process. In the context of early ecological science (such as the work of Frederic Clements), "succession" described the change of a plant community over time until it reached an ideal climax state. While ecologists have since questioned and complicated their understanding of this process, the term has taken on another sense for some agricultural writers, namely

[2] See, for instance, the magazine *Top Producer*.

the well-known agrarian Wendell Berry. For Berry, a "succession" as opposed to a series indicates some kind of meaningful inheritance passed down between bodies, a developmental relationship between generations that marks progress or growth over time rather than just a sequential set of generations, one following after the other (105).

In the phrase "succession planning," succession alludes to successive generations of profitable plant bodies and directly indicates successive generations of profitable human bodies. An example of a succession planning guide for a family farm is provided in *Farm Journal*'s online *Legacy Project,* which is a collection of resources (tools, worksheets, a workbook, meeting agenda guide, and so forth) that encourages farm operators to organize a meeting with their families and plan the future of their businesses, in much the same systemic way that a corporation might outline its strategic plan and goals and for some of the same reasons: peace of mind, maintaining control, maximizing profit. The farmer is guided into the role of an administrator or manager by an economic logic. The headline of a feature story on the main project page by Kevin Spafford advises readers "Don't Put It Off," warning them that the uncontrolled passage of time is dangerous. The article blurb continues, "The decision to plan is smart business. For a farmer, planning for succession is a clear statement that the long-term success of the family operation is important." The interest of more-than-human stakeholders in the world is not relevant; what matters is the farmer's assertion of family values by mitigating risk and creating a written plan for the future.

With its emphasis on family planning, this rhetoric is also connected to the process of human reproduction and the decision of a human couple to have a child or start a family. Planning for a family and for a family farm involve the management of time as well as of a place. This becomes evident in the questionnaire "Are You Ready?", a list of 20 statements that farm operators are asked to assign a number to based on their level of agreement or disagreement. In effect, the questionnaire asks farmers to situate themselves and the land (and plant bodies) that they farm in relationship to a discourse based on reproductive familial bonds ("Maintaining family ownership of the farm/agribusiness is important," "Only lineal descendants should be allowed to own the family agribusiness"), future security ("The owner(s) can retire without converting business equity to cash," "The farming operation provides financial security"), generational seniority

("The senior generation can allow the next generation to make mistakes and learn from experience"), earned property ownership ("Ownership is a privilege, not an entitlement"), corporate management for profit ("The current operation is run like a business with standard operating procedures . . ,"), and American Protestant ethics of work and success based on bodily exertion ("Active family members should receive ownership in proportion to their commitment [blood, sweat and tears]," "The next generation has a strong work ethic").

Some of these statements seem mutually exclusive. For instance, compare "Business success is more important than family harmony" with "Active family members receive adequate compensation for their time, commitment and loyalty to the family operation." Is simply belonging to a family enough to earn a person something, or must family members prove themselves to be successful in business in order to retain their place in the harmonious corporate order of the farm operation? This quandary might be resolved with two more statements, "The family recognizes and acknowledges opposing objectives between active and inactive owners" and "The family communication style is very open and candid." A distinction between active and inactive, or those who are actually working on the farm and those who are not, seems to be something that must be openly admitted when weighing the economic interests of the operation. If this is done, there is the potential that one can achieve both business success and family happiness. Finally, these statements value the creation of a written rather than oral agreement, presumably because of the legal context of inheritance: "The family has shared succession intentions, but not in a written format"; "There is a written plan for operational growth and development."

While these statements aren't offered explicitly as a series of values or core beliefs, the scoring system – in which agreement means the farmer-reader assigns a higher numerical score, up to 5 points per statement – rewards affirmation, to a degree. A very high score is seemingly good but "may also point to over-confidence and/or a lack of empathy for familial tension." Therefore, the best scores are a little lower: the second category is described as having a "healthy respect for the complexity of succession planning, yet maintains a realistic concern for the family's ability to create a positive outcome; the third category also indicates "humility and a realistic expectation"; and the fourth category reaffirms "a need for open discussion among active family members regarding succession intentions," with a recommendation to set up family meetings. The fifth and lowest category doesn't admit any possibility that there might be in the farmer-reader honest disagreement with any of the statements given; instead, "This score points to multiple concerns; the process of succession planning may be premature at this time. Additional assessments, personal counsel and specific plans of action may help to improve the situation." No matter what one believes, succession planning emerges as the answer. The questionnaire cannot be faulted for this, of course. The instructions state that it is designed to help the farmer-reader discern "if you are ready for succession." The questions of what succession should mean, when and how it should happen, and what concerns or values (social or ecological or economic) should inform the process are taken for granted in the audience. Importantly, so is the definition of what a "family" is or could be.

In the previous commercials and planning guide, plant bodies in human discourse have been linked with narratives of generational growth ("generations," "succession," "inheritance," "legacy,") as well as temporal and spatial control ("eradication," "victory," "colonization," "end-of-season clean"). Plant bodies are "producers" whose reproduction must be organized in straight lines in fields, and the bodies of human settler farmers are also "top producers" whose agricultural operations must be economically organized and stabilized through the reproduction of a discourse and value system. In the next and concluding text, it becomes clear how plant bodies are rhetorically and pedagogically linked to the life and health of human bodies in terms of "population" and "civilization." The textbook *Hartmann & Kester's Plant Propagation: Principles and Practices* is now in its eighth edition, and given its longevity and use in land-grant university classrooms by students with particular interest in the reproduction of plants, it merits particular attention with regards to the discursive and material reproduction of plant bodies on the Great Plains.

As a whole, this textbook includes the usual features of the genre (bolded vocabulary words with pull quotes of definitions, diagrams and illustrations of scientific processes, questions for comprehension and discussion, case studies, a clearly articulated system of chapter and section headings, references, indexes, etc.) and covers theoretical and informational background material (about the kinds of propagation being discussed and how they botanically function) as well as practical methods and instructions about how to do work in this field (about how to effectively carry out different kinds of propagation). While there is much of interest (the biological explanations of different sexual and asexual kinds of plant reproduction in particular call for more informed analysis than I can offer here), the opening chapter on "How Plant Propagation Evolved in Human Society" is a useful starting point for the purpose of thinking through the logic of the book. Preceding this chapter and part one in general is this paragraph, on a light green page with the image of a six-petaled flower (or an asterisk) as a faint watermark behind the text:

> This book about **plant propagation** not only describes procedures originating thousands of years ago, but also the application of recent scientific advances. Plant propagation can be described as *the purposeful act of reproducing plants*. It has been practiced for perhaps the past 10,000 years, and its beginning probably marks the start of civilization. The traditional concept of a propagator is a skilled technician who loves plants and who acquired the art from traditional skills learned by experience, or whose knowledge was handed down from one generation to another. Today, propagation may be carried out by an array of general and specialized industries that produce plants to feed the world; to provide fiber, building materials, and pharmaceuticals; and to enhance the world's beauty. (Hartmann et al. 1, original emphasis)

There are several georgic elements to notice here. The first is the valuation of propagation as intentional or "purposeful," and the second are the agents of that purpose. In the past, the propagator was an individual whose skills were gained from first-hand experience or were inherited: skills were produced and reproduced through the individual body and the social body. In the present, the propagator has become an industry – a corporate body. Third, the *civilized* human social body is what emerges from this narrative of bodily actions through time. The beginning of plant propagation "thousands of years ago" marks the beginning of "civilization"; therefore, civilization is implicitly continued through the contemporary environmental, social, and culturally aesthetic missions of industrial plant propagation: "to feed the world; to provide fiber, building materials, and pharmaceuticals; and to enhance the world's beauty."

The first chapter of the book expands on these themes. Though it surveys plant propagation throughout time, it locates the civilized body primarily within the development of Western agriculture, "the deliberate cultivation of crops and animals for use by humans," which allowed hunter-gatherers to settle "at the same site for long periods of time, thus creating centers of activity that eventually would become cities and countries" (2). (Plant propagation is one of the five activities that make up agriculture.) The evolution of agriculture from hunting and gathering results in the production of several new bodies: 1) the economically classed human body, since "Human organization changed from subsistence existence, where everyone participated in the production of food and other items, to a division of labor between agricultural and non-agricultural segments of the population, and even to specialization within the agricultural segment," 2) the domesticated plant body, which represents a human technology-driven advance over the "wild" plant body because "During domestication, crop plants had evolved beyond anything that existed in nature," and 3) the persistently undesirable plant body: "As the fields used to grow plants near human sites were disturbed and became depleted, certain aggressive plant species also were spontaneously established in these sites. These so-called weedy species have become a part of the agricultural system and more or less evolved along with cultivated plants" (3). Alongside the dominant narrative of normalized agricultural plant bodies that produce civilization, then, comes a shadow narrative of non-normative and unnatural plant bodies that "spontaneously" reproduce in a way that unsettles civilization; human selection or influence upon habitat is not implicated in encouraging the reproduction of these plant agents. According to the textbook, weeds are just one of the "side effects that have continued to create problems" (3).

The next section of the chapter outlines the "Organization of Human Societies," in which a variety of civilizations from around the Earth (such as the Middle East, China, and the Americas) are narrowed through time to a mostly Western story, focusing on Greek and Roman societies and medieval societies in Europe. These historical locations contribute to the narrative of capital and civil development described earlier, emphasizing the controlled production of *property*, since in Greek and Roman times "Control of land and agricultural surplus was the key to power and wealth" (4) and in the medieval time, "the specific skills and knowledge of the plant propagator were possessed by specific individuals. These skills, considered 'trade secrets,' were passed from father to son or to specific individuals" (5). Note the remark about lineal, patriarchal inheritance through generations at this time in history; elsewhere in the section, there is no similar attention to how in other times and places, women were the primary plant propagators.

The rest of the chapter continues the temporal narrative of developing the natural spaces of the planet into civilized places, looking at exchanges of plant bodies linked to the spread of civilizing bodies. Readers learn that, "Early movement of useful plants often followed military expansion into different countries when the invading soldier brought plants from his home country into a new land. Conversely, returning soldiers introduced to their homelands new plants they found while on a military campaign" (5). Plant bodies are also collected, preserved, studied, and propagated in new places based on colonial scientific explorations (5-6) and the Morrill Act, which established land-grant institutions of higher education in the United States West (6-8). Finally, the *nursery* becomes an important location for production and reproduction of plant bodies through time (8-11), suggesting another connection to human reproduction and children.

Following a page with a table of contemporary plant propagation societies and organizations (11), the chapter concludes with a brief paragraph under the heading "The Modern Plant Propagation Industry." Readers are taught that it is "large and complex, and involves not only the group that multiplies plants for sale and distribution, but also a large group of industries that provides services, sells the product, is involved in regulation, provides consultation, carries on research, or is involved in teaching" (12). However, students are reassured that there is a place for their laboring bodies as beginning propagators of plant bodies: "The key person in this complex is the plant propagator who possesses the knowledge and skills either to perform or to supervise the essential propagation task for specific plants" (12). The gender of this person is neutral, but his or her actions are firmly planted within contemporary georgic discursive and material practices.

****

Plant bodies produce human bodies by materially constituting them, serving as producers in the food web in which their role is to capture solar energy and transform it into the harvestable calories that allow human bodies to sustain life, to grow, and to reproduce. But plant bodies also are utilized in the production and reproduction of human bodies (for instance, the body politic) through intertwined discourses of ecology, economy, settler colonialism, and sexuality – and the process works both ways. Human bodies materially produce and reproduce desirable plant bodies in agriculture, and discursively produce plant bodies that reproduce desirable human cultural bodies and values.

In the georgic[3] discursive and material practices of the North American Great Plains, what is desired tends to be determined based upon the cultural norms of a Western economic system that values market profitability and the increase of wealth. What is desired also tends to be based upon a goal of feeding human bodies, allowing them to sustain life, grow, and reproduce other human bodies for the future good of the world. In both of these desires, normality is based on controlled development through a length of time, so human bodies are deemed normal when they participate in the cultural logic of valuing "longevity as the most desirable future" (Halberstam 4). Here I quote queer theorist Judith Halberstam, who describes a "middle-class logic of reproductive temporality" that forms lifestyles based on "family time" and the "time of inheritance," the latter of which "connects the family to the historical past of the nation, and glances ahead to connect the family to the future of both familial and national stability" (4, 5).

In *Sowing Empire: Landscape and Colonization*, Jill H. Casid further explains how this powerful logic of normativity is based upon co-reproduction, which involves human and plant bodies in both the discourses and material practices of science, capitalism, imperial colonization, and sexual pleasure and reproduction. The sexual language of planting and transplanting engenders the terms of populations and colonies: "In the scene of the founding imperial gesture of sowing seed, to plant was to make colonies" (xvii). The colonization of land *is* the colonization of the body and the creation of civilized versus wild bodies, both in terms of humans and plants (think of the cultural distinctions between crops and weeds, settlers and savages, natives and

[3] In the context of the United States, the georgic tradition can be located more specifically within a tradition of agrarianism, an umbrella term encompassing philosophies that describe and advocate for the social and economic patterns of agricultural, often rural, lifestyles. Euro-American agrarianism has a long history in the United States, dating back to Thomas Jefferson's vision of a democratic nation of yeoman farmers, and it has developed along with settler colonialism within a capitalist market economy (Donahue 39). Of course, there are also longer-standing alternative "agrarian" and agricultural traditions in Native American cultures.

invasives). Casid traces this narrative's material implications: "Agricultural metaphors of the insemination of woman as furrowed soil were used from Virgil's *Georgics* onward to make heterosexual reproduction and the possession or husbanding of land seem at once synonymous and as preordained and necessary as the cycle of the seasons and the harvest. With this materializing metaphor of the practices of agriculture and landscaping as heterosexual reproduction, to plant was also to produce imperial subjects to populate the colony or work the plantation machine" (xvii). In the settler colonial landscape as well as the imperial landscape, norms are identified, controlled, and reproduced through gendered social and material bodies in order to reproduce the nation and ensure its future. In fact, because settler colonies must reproduce themselves (rather than be resupplied from the imperial center), the dynamic of normatively georgic reproduction is even more important in the settler-colonial situation.

If queer means a gay or homosexual human, or simply a non-heterosexual human, it tends not to be associated with human sexual reproduction (which is often assumed to be heterosexual), though of course there are gay families, heterosexual couples who have homosexual children, bisexual and transgendered people who have or adopt children, and so forth. But if queer more generally means non-normative (when the norm is georgic, settler colonial reproduction), it might lead us to consider more than human bodies and more than human sexual reproduction. Queer in this way gestures toward non-normative plant bodies as well as toward non-normative forms of economic, ecological, cultural, rhetorical, and trans-corporeal reproduction. Importantly, such a definition of queer also gestures toward a more critical perspective on what is discursively and materially presented as "natural" with regard to the reproductive development and future of the nation.[4]

A queer perspective upon reproducing plant bodies on the Great Plains could generate more ways to inquire and intervene into the

---

[4] Note that this theoretical turn is not simply moralistic, and does not necessitate we denounce what is normative as bad and embrace what is non-normative as good in the context of environmental ethics. Instead, the point is to open up critical conversations about what is normative, why, and what the effects of those norms are. For instance, normative family-operated farms and ranches may restore prairie plant communities; conversely, hybrid crops and genetically modified organisms may reproduce in non-normative ways.

contemporary texts I have analyzed. For instance, is family necessarily heteronormative in the succession planning guide, and might there be room for non-lineal, non-reproductive, or non-nationalistic familial relations? Could the history of plant propagation be shared with students in such a way that denaturalizes patriarchal and national institutions of passing down knowledge, and that includes alternative indigenous and feminist ways of knowing and growing plants? Finally, could the battle between masculine technological, agricultural authority and weeds be queered – not just in terms of gender, but also ecologically and economically? If weeds are plant bodies that "obstruct our plans, or our tidy maps of the world," and weeds are "not only plants in the wrong place, but plants which have slipped into the wrong culture," they might offer an opportunity for us to perceive and actively respond to the naturalized, normalized maps of place and culture that structure humans' discursive and material choices – rather than passively reproducing them (Mabey 1, 11).

Weeds might offer a way to decouple or denaturalize the linked concepts of reproduction and growth when it comes to plants. In his chapter "The Body of the Plant," Michael Marder argues that "Metaphysics and capitalist economy are in unmistakable collusion, as they militate against the dispersed multiplicities of human and non-human lives; economic rationality, which currently treats plants as sources of bio-energy or biofuel, converts, concretely and on the global scale, the metaphysical principles of sameness and identity into the modes of production and reproduction of material existence" (55). For Marder, the body of the plant – which grows rhizomatically, bi-directionally from the middle (63), and which can be theorized in such a way that "The leaf usurps the originary status of the seed" (82) – subverts this metaphysical paradigm. He in fact concludes that plants "are the weeds of metaphysics: devalued, unwanted in its carefully cultivated garden, yet growing in-between the classical categories of the thing, the animal, and the human . . . and quietly gaining the upper hand over that which is cherished, tamed, and 'useful'" (90). Might this mean that plants also can grow in between capitalist economies (rather than only reproducing those economies), or that non-useful weedy growth (in contrast to the economically rational reproduction of crop monocultures) might help us imagine less exploitative economic relationships between humans and plants?

Finally, commercials for herbicides to prevent weeds from reproducing can be read in such a way that helps us see the ongoing presence, and disciplining effects, of settler colonialism on the Great Plains. Further analysis along these lines might make visible more ways in which the struggle to reproduce more desirable kinds of *plant* material bodies is discursively represented as necessary for the reproduction of more desirable kinds of *human* social bodies, and suggests ways in which such representations might be contested or challenged.

Like critical plant studies in general, then, a queer way of thinking about reproducing plant bodies calls attention to how plants and humans exist in trans-corporeal relationships. Trans-corporeality can be dangerous, but it can also be sustaining, and in any case, it is unavoidable. If weeds and invasive plant species are a problem on the Great Plains for both farmers and ecologists, there is not persuasive evidence this problem will be solved in academic or popular culture by continuing to think about these plants as savage creatures or aggressive intruders that threaten civilization and need to be combated and controlled. Such an approach serves corporate technology, settler colonialism, and extractive capitalism, and has disturbing ramifications for marginalized human cultures and human communities that are not normalized or naturalized as desirable. We might find better discursive and material solutions by remembering how humans have selected for these reproducing plant bodies, how humans then organized them into the subjective category of "weeds," and how these seemingly non-desirable yet selected-for weedy plants have fed humans, anchored soil, and revealed systemic patterns about the more-than-human world. As Richard Mabey writes, "Weeds thrive in the company of humans. They aren't parasites, because they can exist without us, but we are their natural ecological partners, the species alongside which they do best" (12). This is a queer sense of "natural," and through a queer understanding of reproducing plant bodies on the Great Plains, humans might come to reproduce not what we already know, not what we already are, not what we already want, but something different.

Works Cited

Alaimo, Stacy. *Bodily Natures: Science, Environment, and the Material Self.* Bloomington: Indiana University Press, 2010. Print.

"Are You Ready?" *Farm Journal Legacy Project.* Farm Journal, 2013. Web. 23 May 2013.

Bader Rutter. "Dow AgroSciences PowerFlex – Battleship." Online video clip. *YouTube.* YouTube, 19 March 2010. Web. 23 May 2013.

BayerCropScienceUS. "Capreno Herbicide – 'Man vs Weed.'" Online video clip. *YouTube.* YouTube, 23 May 2012. Web. 23 May 2013.

Berry, Wendell. "A Practical Harmony." *What Are People For?* New York: North Point Press, 1990. 103-108. Print.

Carter, Sarah. *The Importance of Being Monogamous: Marriage and Nation Building in Western Canada to 1915.* Edmonton: University of Alberta Press, 2008. Print.

Casid, Jill H. *Sowing Empire: Landscape and Colonization.* Minneapolis: University of Minnesota Press, 2005. Print.

Donahue, Brian. "The Resettling of America." *The Essential Agrarian Reader: The Future of Culture, Community, and the Land.* Ed. Norman Wirzba. Lexington: University Press of Kentucky, 2003. 34-51. Print.

"FAQs." *Capreno.* Bayer CropScience, 2013. Web. 23 May 2013.

"Glyphosate." *Wikipedia, The Free Encyclopedia.* Wikimedia Foundation, Inc. 22 May 2013. Web. 23 May 2013.

Halberstam, Judith. *In a Queer Time and Place: Transgender Bodies, Subcultural Lives.* New York: New York University Press, 2005. Print.

Hartmann, Hudson T., Dale E. Kester, Fred T. Davies, Jr., and Robert L. Geneve. *Hartmann & Kester's Plant Propagation: Principles and Practices.* 8th ed. Upper Saddle River: Prentice Hall/Pearson, 2011. Print.

"How Does Weed Resistance Develop?" *Monsanto.* Monsanto Company. 2002-20013. Web. 23 May 2013.

Mabey, Richard. *Weeds: In Defense of Nature's Most Unloved Plants.* New York: HarperCollins, 2010. Print.

Marder, Michael. *Plant-Thinking: A Philosophy of Vegetal Life.* New York: Columbia University Press, 2013. Print.

Mortimer-Sandilands, Catriona, and Bruce Erickson. "Introduction: A Genealogy of Queer Ecologies." *Queer Ecologies: Sex, Nature, Politics, Desire*. Ed. by Catriona Mortimer-Sandilands and Bruce Erickson. Bloomington: Indiana University Press, 2010. 1-47. Print.

Scott, James C. *Seeing Like a State: How Certain Schemes to Improve the Human Condition Have Failed*. New Haven: Yale University Press, 1998. Print.

Top Producer. Farm Journal, Inc., 2013. Web. 23 May 2013.

Contributors

Since completing her Ed.M. at Rutgers University and M.A. in English at The College of New Jersey, **Stacey Artman** has been working as a director at the Rutgers Learning Centers and as an instructor of English and pedagogy. Her research focuses on human rights studies and investigates how narrative and metaphor affect – and are affected by – socially influenced interpretations of our worlds and the trauma that occurs within them. Recently, her research has expanded to include ecocritical perspectives and cognitive literary studies in examinations of how humans use the natural world to understand themselves and their experiences of oppression.

Ria Banerjee is a doctoral candidate in English at the Graduate Center, CUNY. Her areas of interest include the literary fin-de-siècle and British modernism, and she has presented at professional conferences on modernists like T. S. Eliot, D. H. Lawrence and R. M. Rilke, and "new woman" writers like Rachilde and George Egerton. Her work is forthcoming in *The D. H. Lawrence Review* and the *Journal of Modernist Periodical Studies* (summer 2012).

Lorraine Burdett is a Ph.D. candidate in the School of Arts and Media at the University of New South Wales. Her dissertation is on Patrick White and Cold War American literary and cinematic culture, specifically focusing on the role of hybridized objects in the production of subjectivity. Her research interests and publications focus on twentieth and twenty first century American fiction and film, surveillance studies and thing theory.

Graham Culbertson, who received his Ph.D. in English at the University of North Carolina at Chapel Hill in 2013, studies nineteenth- and twentieth-century American literature and film. His dissertation examines realist and naturalist narratives of the American city in juxtaposition with the urban narratives found in the contemporaneously emerging disciplines of urban planning, sociology, and reform journalism. He is currently teaching as a post-doctoral fellow at UNC, where he is expanding his previous work on Washington, DC, New York City, and Chicago to encompass the city of Los Angeles and the *roman noir* and *film noir* traditions.

Lynne Feeley is a doctoral student in English at Duke University. She studies eighteenth-century circumatlantic literature and culture. She is writing a dissertation on race and ecology during that period.

Ubaraj Katawal is Assistant Professor of English at Valdosta State University. His articles have appeared in *Postcolonial Text* and *Interdisciplinary Literary Studies: A Journal of Criticism and Theory.*

Hindi Krinsky, Ed.D., is a secondary English Language Arts teacher and college instructor of Secondary Education at Brooklyn College-CUNY. Her most recently published scholarly works include Critical Literacy in English Literature (with Priya Parmar) and contributions to *The Internet and Higher Education*, *Sequential Art Narrative*, and the forthcoming *More Pie: Essays on Entering Academia*. She has presented her work at a variety of national and international conferences.

Aubrey Streit Krug is a Ph.D. student in English at the University of Nebraska-Lincoln, where she studies the Great Plains, American and Canadian literature, ecocriticism, and place-conscious education. She is particularly interested in ethnobotany and in plants in bioregional, agricultural, and Native American literature and rhetoric. Streit Krug earned her B.A. from Bethany College in Lindsborg, Kansas and her M.A. from UNL. Her reviews have been published in *Great Plains Quarterly* and *ISLE: Interdisciplinary Studies in Literature & Environment*, and her scholarship has been published in the *Journal of Ecocriticism*.

Randy Laist is Associate Professor of English at Goodwin College. He is the author of *Technology and Postmodern Subjectivity in Don DeLillo's Novels* and the editor of *Looking for* Lost: *Critical Essays on the Enigmatic Series*. He has also published dozens of articles on literature, film, and pedagogy.

Stephanie Lim is an English graduate student at California State University, Northridge, where she also works as a Teaching Associate and Supplemental Instructor. She enjoys learning about new media, popular culture, and critical theories (specifically, Marxist, Post-Modernist, and Post-Structuralist schools), as well as reading American Literature from the Great Depression on. Although her degree emphasis is Literature, her true passion is musical theatre, and she continually strives to merge musical theatre studies with more traditional, literary studies so as to fully understand the larger, cultural implications of theatrical productions.

Charlotte Pylyser is a Ph.D. candidate in cultural studies at the Catholic University of Leuven. Her research focuses on the contextual (including paratextual) characteristics of the international comics and graphic novel field and of the Flemish graphic novel in particular. She is most notably concerned with the role which the process of institutionalization plays in the (discursively created) emergence and (cultural) function of the Flemish graphic novel. Charlotte sits on the editorial board of *Image [&] Narrative*, a peer-reviewed e-journal on visual narratology and word and image studies.

Betsy Winakur Tontiplaphol is Associate Professor of English at Trinity University in San Antonio, Texas, where she specializes in nineteenth-century British literature and regularly teaches a seminar about Jane Austen. In addition to Austen, she has written about a number of other nineteenth-century literary figures; her first book, *Poetics of Luxury in the Nineteenth Century: Keats, Tennyson, and Hopkins*, was published as part of Ashgate's Nineteenth Century Series in 2011. Her current project examines the relationship between British Romantic poetry and nineteenth-century ballet.

Rhona Trauvitch is a visiting lecturer in Comparative Literature at the University of Massachusetts Amherst. She recently published a chapter, "Alternate History as Countermonument," in *We Are What We Remember: The American Past Through Commemoration* (eds. Laura M. D'Amore and Jeffrey Meriwether, Cambridge Scholars Publishing). Her paper, "Quantum Mechanics and *Mise en Abyme*: The Reader as Observer," has been accepted for inclusion in the edited volume *Interface between Literature and Science: Cross-disciplinary Approaches to Latin American Literature* (ed. Victoria Carpenter, under review by Lehigh University Press). Rhona's research interests include narratology, literature and science, and the Bible as literature.

Akemi Yoshida received her BA from the University of Tokyo, and MAs at both Tokyo and University of Sussex. She is presently an English instructor at Nagoya Institute of Technology, Japan. Her current academic interest lies in the representation of musical sensibility and musically talented characters in the late Victorian novel. Her publications include "A Comparative Reading of Mona Caird's *The Wing of Azrael* and Thomas Hardy's *Tess of the d'Urbervilles*" (2010) and "Voice and Presence in *Callas Forever*" (2013).

Index